BUILDING SECURITY IN THE PERSIAN GULF

ROBERT E. HUNTER

NATIONAL SECURITY RESEARCH DIVISION

This report results from the RAND Corporation's continuing program of self-initiated research. Support for such research is provided, in part, by the generosity of RAND's donors and by the fees earned on client-funded research.

Library of Congress Cataloging-in-Publication Data

Hunter, Robert Edwards, 1940-
Building security in the Persian Gulf / Robert E. Hunter.
p. cm.
Includes bibliographical references.
ISBN 978-0-8330-4918-6 (pbk. : alk. paper)
1. Security, International—Persian Gulf Region. 2. Persian Gulf Region—Strategic aspects. I. Title.

JZ6009.P35H86 2010
355'.0335536—dc22

2010017767

Cover photo: NASA, Visible Earth (http://visibleearth.nasa.gov/)

Published 2010 by the RAND Corporation
1776 Main Street, P.O. Box 2138, Santa Monica, CA 90407-2138
1200 South Hayes Street, Arlington, VA 22202-5050
4570 Fifth Avenue, Suite 600, Pittsburgh, PA 15213-2665
RAND URL: http://www.rand.org/
To order RAND documents or to obtain additional information, contact
Distribution Services: Telephone: (310) 451-7002;
Fax: (310) 451-6915; Email: order@rand.org

Preface

The United States is now in the process of reducing its force presence in Iraq, and there is a December 31, 2011, deadline for final withdrawal. This process is being very carefully planned and executed, and considerable effort and analysis have gone into it.[1] Less well considered and planned is what happens afterward, not just in Iraq but in the region of the Persian Gulf, particularly in regard to regional security. This is not only shortsighted: It could have a serious impact not just on the region but also on major U.S. (and allied) interests in the Persian Gulf and vicinity. It is now well understood that, when the U.S.-led Coalition invaded Iraq in March 2003, not enough thought and planning had gone into what would need to be done after the conflict. With this experience as background, it is important that the same not happen again regarding other critical issues in the region.

Many important questions will need to be answered about what should happen next. This work addresses several such questions that are related not to specific aspects of the endgame within Iraq itself but rather in the surrounding region. At the same time, rather than concentrating on immediate tactical questions, it primarily takes a longer view, focusing not so much on the drawdown of U.S. and Coalition forces as on what could be done in the medium to long terms to help foster security in the Persian Gulf region as a whole. Thus, in addition to considering some short-term factors, this work develops criteria and parameters for *a new security structure for the region of the Persian Gulf.*

This work has two primary goals. The first is to determine and develop means for increasing the chance of achieving long-term security within the Persian Gulf and environs (as far west as the Levant and as far east as Iran and, in some regards, Afghanistan and Pakistan). The second, within the context of fully securing U.S. interests and those of its allies and partners, is to reduce the long-term burdens imposed on the United States in terms of (1) military engagement and the financial cost of providing security; (2) risk, including to U.S. forces; and (3) opportunity costs, especially in relation to East and South Asia, the Russian Federation, and management of the global economy.

[1] For an assessment conducted outside the U.S. Department of Defense on the process of withdrawing forces from Iraq, see Perry et al., 2009.

Furthermore, *such a security structure could serve as a sort of firebreak* to help prevent normal stresses and strains from escalating to a level of tension or conflict at which none of the parties gains and all may lose. This is a cost-benefit analysis that concludes that, in the common interest, certain behavior should be ruled out of bounds to prevent it from spiraling out of control.

In developing such a new security structure, it will of course be necessary to factor in developments in Afghanistan and Pakistan, given that what happens there will have a significant impact not just on Iran but also both on other countries in and concerned with the Persian Gulf and on overall U.S. interests in and perspectives on the entire region of the Middle East and Southwest Asia. The entire region must be dealt with in terms of all its interconnections—*holistically*, as it were.

The term *structure* is used here advisedly. The alternative is to consider only a *system* of security, which is the sum of all the factors that describe the existential quality of security—say, within a region—however these factors have developed and whatever they are. The idea of structure, by contrast, connotes conscious effort to develop a framework, some aspects of which can be formal and some informal. This framework will help (1) to channel and concretize expectations, (2) to enable different parties to the structure to calculate their own self-interest in relationship to practices set up by the structure, and (3) to enable these parties to see whether taking part will bring them more advantages than not doing so. A further merit of looking at a structure rather than just a system for security is that doing so permits the development of patterns of behavior and expectations that can endure. A security structure, as opposed to simply a security system, implies the existence of rules—or at least approaches—that are written down; functional arrangements; institutions, whether rudimentary or more developed; consultative mechanisms that all participants respect to a degree sufficient for the mechanisms to work for all; and processes for dealing with threats and challenges to security (and agreed criteria for determining what they are) within the region that the structure encompasses.

What follows is an analysis of many, but by no means all, of the factors in the region of the Persian Gulf and environs that arise in regard to developing security in the broadest sense of the term. The analysis is designed to begin a process of developing means for providing security for the region, especially because the events of the last several years, including the wars in Iraq and Afghanistan, will, perforce, create a new *system* of security that is significantly different from that of the past. That bridge has already been crossed. The question is rather *what* will fill the place that the collapse of the old system of security—as shaky as it was—has created.

Of course, in this particular region, development of a comprehensive and effective security *structure* may not be possible: Differences among states and religious and other groupings may prove to be simply too deep and pervasive. But that has also been true in other parts of the world when efforts first were made. The hope is that rational calculus of the best means of advancing individual interests will lead to acceptance that at least some forms of structured cooperation related to security will prove more worthwhile to all concerned than a state of anarchy. What follows is, thus, an effort to spell out some of the key factors in getting from here to there. Political will is of course needed to turn vision and analysis into reality.

This work does not deal with all of the practical details that would have to be worked out. For example, it does not analyze in detail what would be involved, in practice, with regard to the deployment of U.S. military forces, or those of other countries, in or near the Persian Gulf for the long term (much less their composition and size). These and other implementation issues will need to be investigated in a next stage, building on the analysis and recommendations in this work. This work proposes a framework within which detailed issues can be dealt with.

This report results from the RAND Corporation's continuing program of self-initiated research. Support for such research is provided, in part, by the generosity of RAND's donors and by the fees earned on client-funded research.

This research was conducted within the RAND National Security Research Division (NSRD) of the RAND Corporation. NSRD conducts research and analysis for the Office of the Secretary of Defense, the Joint Staff, the Unified Combatant Commands, the defense agencies, the Navy, the Marine Corps, the Intelligence Community, allied foreign governments, and foundations.

For more information on the RAND National Security Research Division, contact the Director of Operations, Nurith Berstein. She can be reached by email at Nurith_Berstein@rand.org; by phone at 703-413-1100, extension 5469; or by mail at RAND, 1200 South Hayes Street, Arlington VA 22202-5050. More information about the RAND Corporation is available at www.rand.org.

Contents

Figures

Summary

With the winding down of the U.S. military engagement in Iraq and with other developments in the region of the Persian Gulf, the United States, along with its allies and friends, faces the need to define a long-term strategy for the region. Part of that strategy will relate to immediate issues of continuing U.S. involvement in Iraq, part to challenges posed by Iran, part to developments in nearby Southwest Asia (i.e., Afghanistan and Pakistan), part to the continuing Arab-Israeli conflict, and part to the overall U.S. position in the region and its reputation for "getting right" its strategy and approach to the region and the steps taken to implement them.

The nature of U.S. and Western interests in the region—e.g., the secure export of energy, stability and predictability, counterterrorism, relations with other great powers, geopolitics and geoeconomics in general—means that the United States will have no choice but to remain a deeply engaged power in the region. Of course, the terms, conditions, qualities, dimensions, and application of that power and related influence are subject to debate, decision, and responses to events that have not yet happened or, in some cases, even been imagined. But the United States' continued, indeed permanent, engagement in the region has already been determined by its interests.

There are many elements to consider and questions to be answered. This work focuses on two. The first is the best strategy and approach to promote medium- to long-term security and stability in the region as a whole, consonant with the basic interests of the United States, its allies and partners, and the interests of regional countries that are prepared to value the reduction of both tensions and the risk of conflict more than pursuing national ambitions that are divisive and could lead to conflict.

The second element considered here is based on two premises: First, for the United States to continue playing a major role in the region's security, that role must be grounded in strong popular support at home; second, there is, thus, value in finding means to promote regional security and stability at a lower cost than is now being exacted from the United States in terms of blood, treasure, and opportunity costs. If these premises are correct, then canvassing and analyzing such means is worthwhile, indeed, indispensable.

This analysis focuses on setting parameters for a regional security structure that is designed to have a high degree of likelihood of covering the key factors in play. Eight basic, region-specific sets of parameters need to be considered in formulating a new security structure for the Persian Gulf region: the future of Iraq; Iran; asymmetric threats; regional reassurance; the Arab-Israeli conflict; regional tensions, crises, and conflicts; the roles of other external actors; and arms control and confidence-building measures.

The Future of Iraq

The U.S. drawdown of forces from Iraq would be facilitated if the foundations were being laid for a new security structure for the Persian Gulf, thus firmly placing U.S. policy toward Iraq in a regionwide and long-term context. Those foundations would then provide the basis for devising a more-elaborated structure able to deal with a broad range of Persian Gulf security requirements. A first-step structure, based on short-term needs in Iraq, should include the following elements:

- diplomacy, including a regional conference or conferences that would bring together all the parties relevant to Iraq; a commitment from all parties to the basic goals for Iraq, which are based on its independence, sovereignty, and mastery in its own house, plus an agreement not to use force to affect developments there
- participation in multilateral diplomacy by the United States, Iraq, Iran, Turkey, and the Gulf Cooperation Council (GCC) states, along with the United Nations (UN) and other countries and international bodies by common agreement
- a joint commitment from all regional parties to oppose terrorism in Iraq in all forms and from all sources
- the creation of a Standing Military Commission, composed of all the core members of multilateral diplomacy relevant to Iraq and run under Iraqi leadership, to agree on definitions regarding the long-term military situation in Iraq, to develop limitations on outside involvement, to create a system of inspections, and to devise confidence-building measures (CBMs)
- the creation of a Standing Political Commission, with the same membership as the Standing Military Commission and also run under Iraqi leadership, to clarify the interests of all parties, to build political confidence in one another's activities, to develop tools for dealing with terrorism, and to create expert teams to assess Iraqi material requirements (e.g., reconstruction and development), all as a prelude to future donors' conferences
- development of the concept of the Standing Military Commission and the Standing Political Commission for use in a regionwide security structure.

Iran

Iran is currently the most-important country in the region in terms of the future of Persian Gulf security and stability. For many years, the United States has operated on the assumption that Iran will be largely uncooperative. Recently, this assumption has been reinforced both by the continuing standoff over Iran's nuclear program and by a compound of Iranian statements and actions. Yet, this assumption is worth reexamining. On the one hand, Iran is obviously pleased to see the United States reduce its presence in Iraq, not least because of the proximity of U.S. military forces to Iranian territory. It might, thus, calculate that now is an ideal time to try increasing its influence in Iraq, in part to pursue overall regional ambitions. On the other hand, Iran also needs to be wary of an Iraq that is in turmoil, and it might be willing to assist in stabilizing that country. Which perspective will prevail will likely be clarified during 2010, and it is also likely to be affected by what else is happening—or not happening—in U.S.-Iranian relations.

There is less uncertainty about the potential for Iranian cooperation, whether tacit or explicit, with the United States and other members of the International Security Assistance Force (ISAF) in Afghanistan. There is some clear correspondence of interests between the two parties, including opposition both to the return to power of the Taliban and to the capacity of Al Qaeda to operate from Afghan (or Pakistani) sanctuaries. How compatible interests might translate into positive cooperation is another matter, however.

At the same time, for most Arab states of the Persian Gulf, a key criterion for judging the U.S. drawdown from Iraq and other U.S. policies in the region is whether the United States remains willing and able to thwart Iranian ambitions, not just vis-à-vis Iraq but also in the region of the Persian Gulf in general.

The apposite point for this work is whether Iran would be prepared to play a constructive part in a regional security structure. It has, for some time, been willing to consider security arrangements with regional countries, and it has even proposed a broad-ranging structure, but with the proviso that the structure be composed solely of regional states. In the end, the issue is whether Iran would indeed be prepared to take part in a regional security structure that did involve outsiders, especially the United States and its allies and partners. And, if such a structure were created—that is, if it were not based on the existence of a hostile "other" but rather on promoting security for all—would Iran see its interests as being better served by being inside or outside?

To be eligible to take part in a security structure, Iran would also have to meet some U.S. criteria, including the resolution of the nuclear issue; Iranian abstention from troublemaking in Iraq; Iranian cooperation in Afghanistan (reciprocated by the United States and the North Atlantic Treaty Organization [NATO]); and Iranian abandonment of both support for terrorism and opposition to Arab-Israeli peacemaking. By the same token, Iran would no doubt have its own list of requirements, which could include security guarantees in exchange for meeting the West's requirements, the lifting of all economic sanctions, an end to efforts to destabilize the regime or dismember Iran, some role in both Iraq and Afghanistan, and recognition of a significant role for Iran in the Persian Gulf.

These two lists need not be incompatible. They should be the basis for negotiation, and those negotiations, to have a chance of being effective, need to be comprehensive and inclusive of all issues. Even if negotiations succeeded—a difficult achievement—there would still be value in a regional security structure that would serve as both a firebreak against uncertainties and a means of dealing with inevitable tensions and disagreements. There would also be value in starting with limited Western cooperation with Iran to test the possibilities—a *holistic approach conducted step by step*. Afghanistan is the obvious place to begin such cooperation.

Asymmetric Threats

Some of the most-significant players in terms of regional security are not states but nonstate actors, especially such terrorist groups as Al Qaeda but also other groups that challenge authority through extralegal means or could use violence to forestall the emergence of a viable security structure. A generic term for conflict involving relatively low-cost tools with a potentially high pay-off is *asymmetric warfare*, which is a limited form of *force equalizer*.

Asymmetric warfare, as practiced by insurgent or terrorist groups, has three basic objectives. First is the appeal to hearts and minds through either positive or negative tactics. Second is the use of relatively low-cost instruments (e.g., improvised explosive devices) for tactical gains against relatively high-cost instruments or classic military tactics. Third is the effort to cause political change on the part of the government under attack—ideally, its overthrow—or to affect the politics and the policies of external states that are supporting the government.

The issue of asymmetric warfare is relevant to a new security structure in the Persian Gulf in at least three ways. First is the question whether all participants will foreswear asymmetric warfare. Iran is the key focus of this question in terms of states, but Al Qaeda in Iraq, the Partiya Karkerên Kurdistan [Kurdistan Workers' Party] (operating against Turkey), and some individuals and groups in Saudi Arabia are also primary concerns. Second, there needs to be agreement, including a formal antiterrorism compact that applies also to the Levant, to oppose any use of asymmetric warfare. Third, efforts by local states to oppose asymmetric warfare are needed if they expect the sustained involvement of external states, especially the United States and European countries. Local governments cannot turn a blind eye to asymmetric warfare in their midst and expect the United States and others to be engaged in promoting security and stability.

Regional Reassurance

The development of a new security structure for the Persian Gulf cannot just be an excuse for a reduction in U.S. involvement in the region. It must also account for expectations on the part of regional states and others about the future role of the United States in the region, whether it acts on its own or in league with European and other allies and partners. There are five key elements to providing reassurance:

- The United States should continue conducting the withdrawal and repositioning of its forces from Iraq within a valid strategic framework that accounts for its own interests and those of key regional and extraregional countries rather than in a way that could be represented as war weariness. Indeed, developments in Iraq could require the United States to slow, stop, or reverse its force drawdown.
- The United States can continue to bolster the defenses of regional Arab states in the face of concerns about Iran's development of ballistic missiles, its nuclear program, and its intentions in the region. But care needs to be taken to calibrate the implementation of this policy so that it does not itself further stimulate a potentially uncontrollable regional arms race and preclude possible Iranian participation in a regional security structure at some point.
- The United States could reposition its military forces to provide reassurance to regional countries, but, in so doing, it should avoid areas where the sheer presence of U.S. forces could have a lightning-rod effect of negative popular reaction. It should also avoid adopting force types, quantities, and configurations that could prove destabilizing. There would be great value in communicating to all parties what the United States is doing and in consulting with others to avoid misunderstandings. Further, any challenges to regional states from Iran are less likely to be made in military than in nonmilitary terms. Thus, local countries need to rely more on their own devices, including control of migration and investment flows, unless matters get so out of hand that the United States is pressed to become involved. Regional states will also need to judge whether "security" can be promoted by the willingness and ability of the United States and other Western countries to forge workable relationships with Tehran, an avenue that some regional countries are themselves already exploring.
- The United States could consider providing formal security guarantees to regional states against aggression from their neighbors and, potentially, against threats from sources external to the region. It will also need to continue playing a major role in countering terrorism in the region.
- Through diplomacy, the United States needs to foster among others a positive appreciation of what it is doing and is prepared to do in the future; in some cases, it may need to take steps to produce a deterrent effect in regard to potentially hostile countries. Such efforts could include promoting development of a regional

security structure, provided that it is not characterized as an effort by the United States to gain political dominance in the region or, by contrast, to wash its hands of the region's problems.

The Arab-Israeli Conflict

Calculations about Middle Eastern politics and U.S. engagement in the region continually return to the Arab-Israeli conflict and, more particularly, to Israeli-Palestinian relations. Conventional wisdom holds that continued lack of resolution of the Israeli-Palestinian conflict inhibits the United States' pursuit of its interests elsewhere in the region, including the creation of viable security arrangements. Conventional wisdom also holds that the United States' European allies expect Washington to take an active role in pushing this conflict *toward* conclusion—if not pressing it *to* a full conclusion—in return for their cooperation elsewhere in the region.

But how accurate is the conventional wisdom? If asked, virtually everyone in the Arab/Muslim states in the region and in Europe asserts that the United States must play an active, committed role and drive the process to closure. But this is not the same as a *requirement.* Nevertheless, U.S. efforts in the Middle East, overall, are clearly affected negatively by popular attitudes in the Muslim states of the region regarding the U.S. role in the Palestinian issue. Thus, for all these reasons, U.S. efforts in the Middle East as a whole would certainly *not be made more difficult* if the Israeli-Palestinian conflict were settled. The United States is also likely to gain standing in the region and with allies if it is actively and seriously engaged in seeking that end, as President Barack Obama is now doing. Further, issues relating to Israel's security and to Iran's role in the Middle East (especially its nuclear program) would, unless dealt with effectively, vastly complicate any U.S. efforts to foster a new security system for the Persian Gulf. In at least this element, therefore, there is a clear linkage between the zone of Arab-Israeli conflict and the Persian Gulf and Southwest Asia.

In Israeli-Palestinian peacemaking, however, the devil rests less in the details than in the nature, character, and development of underlying politics and societies. Many Israelis are still not willing to take the classic risks associated with achieving peace. It is also far from clear that there is a valid partner for peace for Israel; indeed, finding one is unlikely to be possible until Gaza ceases to be largely ignored and is instead provided the massive external economic and humanitarian support needed to begin weakening the hold of Hamas. Furthermore, in its current isolation, Gaza is fertile ground for Islamist terrorism's recruiters. At the same time, leading Arab states are waiting to see whether Israel will be prepared to take both substantive and symbolic steps—notably, on its settlements policy—and whether the U.S. president is prepared to run his own risks for peace, denominated, in part, in terms of U.S. politics. For his part, President Obama cannot succeed in pressing serious peace efforts beyond a limited point without prior progress within Israeli and Palestinian politics and society, including an end to Gaza's isolation. Nor will he be able to count on support from Arab states until he shows his own commitment to move forward, not just on tactical issues, such as Israeli settlements, but on the big issues that must eventually be resolved.

The interplay of all these factors argues that the default option should be active U.S. diplomatic engagement in trying to resolve the Israeli-Palestinian conflict and concurrent efforts to reduce threats—to countries in the Persian Gulf region and to others, such as Israel—and to build the basis for lasting security in the Persian Gulf.

Regional Tensions, Crises, and Conflicts

Any viable security structure also has to take account of tensions, crises, and the possibility of conflict, including destabilization, between its members. Both Turkey and Iran have concerns on the last-named score. Meanwhile, there have been serious tensions in Qatari-Saudi relations; indeed, such tensions may help to explain why Qatar welcomes the U.S. military presence in that country. Also, relations between Saudi Arabia and Bahrain have not always been cordial, and neither have those between Saudi Arabia and Yemen. There are also occasional stresses in relations between Saudi Arabia and the United Arab Emirates. The Iraqi government keeps a wary lookout for potential interference from various directions, including some of its Arab neighbors. Relations among GCC members could also deteriorate for one or another reason.

Some Persian Gulf governments considering whether to join a formal security structure could also seek support against *internal* political change. But it is one thing to require that neighbors pledge not to engage in or tolerate subversion and another to require that each member of a regional security structure assist another whose government is being challenged from within. Trying to write that provision into a treaty could put more weight on the arrangements than they can bear.

The Roles of Other External Actors

Since the onset of serious difficulties following the 2003 invasion of Iraq, the United States has already moved significantly in the direction of seeking support from allies and partners in the Middle East.

The Europeans

After the 2003 invasion, the United States had no choice but to take the lead in devising some alternative to the old structure of regional security. Since then, however, it has moved toward seeking support from allies and partners in the Middle East. Most of these allies and partners now accept that they cannot stand aloof from what happens in the region. Not all see eye to eye with the United States on the *nature* of the challenges or on the potential *remedies*, but there is broad agreement that the United States cannot be expected to bear a vastly disproportionate share of the responsibility to act. This is part of an emerging bargain: *The United States will continue to be deeply engaged in European security, but it needs support from its allies in the Middle East and Southwest Asia*, as far east as Afghanistan and even Pakistan.

Furthermore, in many parts of the Persian Gulf region, some Western countries can be more effective than the United States because they do not carry the political baggage of being the legatee of others' colonialism, the invader of a regional country, or a strong supporter of Israel. European states are also well placed to train local security personnel and to provide nonmilitary support. The United States should continue to encourage the involvement of European governments, even at the price of ceding some primacy, sharing political influence, and accepting joint decisionmaking.

Other Key External Powers

The potential or actual engagement of other external powers in the Persian Gulf cannot be ignored, especially because of the region's hydrocarbons. But do any of these actors need to be involved in a new security structure for the region? India and China are developing both interests and ambitions, but it is not clear how much either could contribute or, by contrast, whether they would have the incentives or the ability to confound others' arrangements. They stand to gain assurances about hydrocarbon flow as a free good. Decisions about their potential roles can, thus, probably be made somewhat later.

Russia is a different case. It already plays roles in the Middle East, notably in the so-called Quartet for Arab-Israeli peacemaking. It has involved itself in the issue of U.S. supply routes to Afghanistan, and it has a strong interest in the transportation of hydrocarbons from Central Asia and the Transcaucasus. It is seeking, in general, to return to the ranks of great powers, which implies increased engagement in issues of the Middle East and Southwest Asia. Perhaps most consequentially, Russia has been part of ongoing negotiations regarding the future of Iran's nuclear programs. No doubt, however, it has its own interests regarding Iran and the region as a whole.

Thus, should Russia be invited to join a new security structure for the Persian Gulf? A Russia that is passive but supportive, as it has tended to be in the Quartet, should be welcomed. But Russia's own attitude will depend in part on Iran's stance toward a regional security structure. If Tehran were obdurate, this could work to Moscow's advantage. But if Tehran were positive, then Moscow's role would be diminished. The bottom line is that the creation of a regional security structure should include exploring possibilities with Russia and trying to create incentives for it to be cooperative.

Building Blocks for a Regional Security Structure

In a security structure for the Persian Gulf, there can be value in creating formal political and security commitments among its members. One common form is *collective security*, which is an "all-the-rest-against-one" approach in which all members agree to support one another against a possible threat from one or more of the members (an example of such an organization was the League of Nations). Another popular form of security commitment is *collective defense* (an example is NATO), a form in which all the parties agree to come to the aid of any alliance member threatened from outside: an "all-for-one-and-one-for-all" approach. This work makes no a priori judgment about the need for treaty commitments. Instead, it takes a building-block approach, analyzing possible alternative approaches to security—while setting parameters—that could be drawn upon for a viable security structure. Holding off on making politically binding mutual security commitments until functional arrangements are developed can have two benefits: It can preserve flexibility regarding future membership (e.g., for an Iran that is not part of initial efforts) and it can avoid problems that could arise if regional states are wary, at least at first, of making formal commitments to one another.

Potential Models or Partners

In assessing possible building blocks for a regional security structure for the Persian Gulf, it is worth canvassing alternatives based on experience elsewhere. These potential partners or models are discussed in this section.

Determining what can be done to build a security structure for the Persian Gulf will have to take account of the region's unique features. Like all other security structures in different parts of the world, this one will have to be sui generis. Nevertheless, experience elsewhere can be instructive and offer precedent or practice to inform efforts within the region. One distinction can be made between models of other cooperative efforts or institutions that are based solely on regional states and those that involve outsiders. There are advantages to the former, especially in cases where, as in a number of countries in the Persian Gulf region, the involvement of outside military forces could have a negative effect. By contrast, there can be circumstances in which the involvement of outsiders can have a reassuring quality by demonstrating to local states that their security matters to, say, the United States, and that they would be backed up when need be.

The following are summaries of possible *models* or *partners* for regional states in developing a regional security structure:

- *NATO involvement.* NATO is already involved in the Persian Gulf region via both the NATO Training Mission–Iraq, which will continue for the foreseeable future, and its leadership of ISAF. For some aspects of a Persian Gulf regional security structure, that experience could be apt. NATO's Istanbul Cooperation

Initiative (ICI) includes four of the six members of the GCC—Saudi Arabia and Oman do not participate—and offers NATO support in functional areas, notably training. Allied Command Transformation also offers significant possibilities for cooperation, as do NATO facilities made available to non-NATO militaries for training and similar purposes.

- *The NATO model.* The roles of two NATO institutions might be instructive in the Persian Gulf region. Partnership for Peace (PFP) has proved to be highly effective in enabling states that emerged from the wreckage of the Soviet Union, the Warsaw Pact, and Yugoslavia to develop the tools of military modernization and cooperation. A key function of PFP has been to help damp down tensions and differences between different members. The Euro-Atlantic Partnership Council now provides a forum in which any of these countries can raise their security concerns, especially concerns about each other. Another NATO effort, the coordination of measures for disaster relief through the Senior Civil Emergency Planning Committee, could also provide a model for cooperation among Persian Gulf states.
- *European Union (EU) involvement.* In certain circumstances, the EU could make available its crisis-management capabilities (through its Common Foreign and Security Policy [CFSP]) and even limited military engagement (through its European—now *Common*—Security and Defence Policy [ESDP/CSDP]), although these capacities are very limited. One virtue of EU involvement would be that it is "not NATO" and "not the United States"—facts that might prove more politically appealing to some Persian Gulf states.
- *The EU model.* Despite the formal creation of a Gulf Common Market in 2008, anything like the EU in the Persian Gulf is decades away from being established. However, there are lessons to be learned from European experience in terms of conflict- and tension-reduction through economic cooperation. The CFSP and CSDP experiences meld different approaches to or phases of security into a "one-stop-shopping" model.
- *A Conference on Security and Co-operation for the Persian Gulf (CSCPG).* This organization could be patterned on the Conference on Security and Co-operation in Europe (CSCE), which was designed to help reduce tensions during the Cold War, when different countries wanted to explore ways of cooperating even during a time of basic geopolitical and ideological conflict. Indeed, CSCE proved to be a critical part of the process of bringing the Cold War to an end. At the very least, the "Basket One" aspects of CSCE—basically, security cooperation—might be usefully adapted to a CSCPG in making possible a range of relations between Iran and the Arab states of the Persian Gulf without any of the local countries having first to cede basic approaches or to compromise their interests.
- *An Association of Persian Gulf Nations.* This organization would be patterned on the Association of Southeast Asian Nations (ASEAN), with its ten member states,

and on ASEAN's Treaty of Amity and Cooperation in Southeast Asia, which incorporates 17 non-ASEAN members, including the United States, Russia, China, and the EU. The virtues of this model are that it has developed slowly; it includes tension-reduction mechanisms; it does not require that all disputes be resolved before countries can join; and it relates politics, economics, and security together. The treaty also provides a role for external countries if and when regional countries see this role as beneficial.
- *The Organization of the Islamic Conference (OIC).* All potential regional members of a Persian Gulf security structure belong to the OIC, which includes a rudimentary mechanism for the peaceful settlement of disputes and a convention for combating international terrorism. There are elements of OIC policy and practice that could be useful in developing a security structure for the Persian Gulf.

Arms Control and Confidence-Building Measures

The development of a new security structure for the Persian Gulf region should include both arms control and CBMs. These can help introduce more rationality into the process of determining security requirements and could lead different parties to adjust their policies. Indeed, the impact that regulating military relationships can have on political relationships should not be underestimated. Step one is to keep the military dimension from dominating the politics. An example of a way to do this is to recognize the need to prevent the emergence of a balance of forces that is inherently instable. This is particularly true where weapons of mass destruction are involved.

The following are arms-control measures and CBMs that could be useful parts of a Persian Gulf security structure:

- *multilateral political and military commissions*, whether they include outside powers, such as the United States and European states, or are limited to regional powers. These commissions, patterned on the U.S.-Soviet Standing Consultative Commission created by the 1972 Anti-Ballistic Missile Treaty, would develop techniques for fostering stability.
- *an incidents-at-sea agreement* patterned on that between the United States and the Soviet Union in 1972
- *a freedom-of-shipping agreement* designed especially to create additional confidence in shipping through the Strait of Hormuz, which is important to all littoral states
- *counterpiracy cooperation*, which would be in the common interest of all Persian Gulf maritime nations in countering the rising phenomenon of piracy in the Red Sea and environs
- *a counterterrorism compact*, including practical efforts and cooperation to defeat terrorism

- *a weapon catalog as a prelude to arms control*, which would provide a basis for calculating military balances of power, an important element of a security structure
- *limitations on sales and supplies of weapons* that could be destabilizing. These limitations would be accepted by regional states and established in cooperation with supplier states.

Summary of Recommendations

Important criteria for an effective regional security structure include the following:

- a critical mass of regional states that see taking part in the structure as more likely to provide security than abstaining from or working against it
- willingness to pursue arms-control measures and CBMs, including
  - establishing multilateral political and military commissions to reduce tensions and the risk of conflict
  - creating CSCE-like arrangements (i.e., a CSCPG)
  - developing PFP-like relations among states
  - establishing an incidents-at-sea agreement, a freedom-of-shipping agreement, counterpiracy cooperation, and a counterterrorism compact; cataloging (and defining) military capabilities; and limiting the acquisition of weapons in the region
  - adopting nonmilitary (especially economic) cooperation and integrating military and nonmilitary approaches to security
  - creating means to limit or resolve intraregional squabbles, tensions, and crises
- the integration of regional security efforts within a formal UN mandate to create a rule-of-law basis for cooperation
- the creation of a security structure premised either on universal membership *or*, if a "hostile" state is in the mix, on pursuing containment (but with the possibility of universal membership in the future)
- roles for outside institutions, notably NATO (e.g., through the ICI) and the EU (e.g., through an enlarged EU Mediterranean Initiative)
- a method of dealing with asymmetric threats posed either by member states on their own or with external partners (e.g., such countries as the United States or such institutions as NATO and the EU)
- roles (if any) for Russia, China, and India
- possible security pledges by outsiders, perhaps including deployed forces or other demonstrations of presence and commitment
- outsider involvement in training and other security-support roles
- progressive, rather than one-time, development of the security structure, including a series of regional security conferences
- long-term efforts aimed at internal social, economic, and political development.

Acknowledgments

This project began in 2007 as an effort to bring to bear the thinking and experience of a number of RAND experts on a single topic of considerable importance to the United States: the need for an overarching concept for the security of the Persian Gulf region, particularly following the end of the war in Iraq. The author would like to thank RAND President James Thomson, who supported the development of a project plan designed to make cooperative use of talents across the spectrum of RAND's capabilities. The author of this monograph recruited several RAND senior staff members from diverse issue backgrounds, all of whom have extensive experience both in research and in policy formulation. These were David Aaron, James Dobbins, Stephen Hosmer, Stuart Johnson, Seth Jones, Stephen Larrabee, Thomas McNaugher, Charles Nemfakos, and Peter Wilson. Theodore Karazik and Ghassan Schbley also played invaluable roles, conducting research and interacting with experts in the Persian Gulf region. The author expresses his thanks and appreciation to them all for their invaluable insights in helping to shape the basic framework of this monograph, to suggest key ideas, and to provide good judgment in pointing the study in useful directions.

In the course of research, the author travelled to Western Europe and the United Arab Emirates; consulted widely with experts and government officials—too numerous to name individually—in the United States, Europe, and the Persian Gulf; and drew on other RAND research. He was aided immeasurably by excellent comments and suggestions on drafts of the manuscript provided by David Aaron, Director of the RAND Center for Middle East Public Policy; James Dobbins, Director, RAND International Security and Defense Policy Center; and the two formal reviewers, Gary Sick of the School of International and Public Affairs at Columbia University and former Senior Staff Member for the Middle East (especially Iran) at the National Security Council; and Frederic Wehrey, Adjunct Senior Policy Analyst at RAND and a noted expert on the Persian Gulf region. The author also thanks his assistant, Sabrina Hayes, and, from Publications and Creative Services, Carol Earnest, Erin-Elizabeth Johnson, Steve Kistler, and Jocelyn Lofstrom. Most of all, the author expresses his deep appreciation and thanks to his spouse, Shireen Hunter, who provided inspiration, guidance, and unfailing patience and support throughout the conduct of this project.

Any errors of fact or lapses of judgment in this work are, of course, those of the author alone.

Abbreviations

ACT	Allied Command Transformation [NATO]
ARF	ASEAN Regional Forum
ASEAN	Association of Southeast Asian Nations
C4ISR	command, control, communications, computers, intelligence, surveillance, and reconnaissance
CBM	confidence-building measure
CENTO	Central Treaty Organization
CFSP	Common Foreign and Security Policy [EU]
CSCE	Conference on Security and Co-operation in Europe
CSCME	Conference on Security and Co-operation in the Middle East [proposed]
CSCPG	Conference on Security and Co-operation for the Persian Gulf [proposed]
CSDP	Common Security and Defence Policy
CSTO	Collective Security Treaty Organisation
EAPC	Euro-Atlantic Partnership Council [NATO]
EEC	European Economic Community
ESDP	European Security and Defence Policy [EU]
EU	European Union
EU3	the United Kingdom, France, and Germany
G8	Group of Eight

G20	Group of Twenty Finance Ministers and Central Bank Governors
GCC	Gulf Cooperation Council
IAEA	International Atomic Energy Agency
ICBM	intercontinental ballistic missile
ICI	Istanbul Cooperation Initiative [NATO]
ISAF	International Security Assistance Force [Afghanistan]
JFCOM	U.S. Joint Forces Command
MD	Mediterranean Dialogue
MEK	Mujahedin-e-Khalq
NATO	North Atlantic Treaty Organization
NGO	nongovernmental organization
NPT	Nuclear Non-Proliferation Treaty
NTM-I	NATO Training Mission–Iraq
OIC	Organisation of the Islamic Conference
OPEC	Organization of Petroleum Exporting Countries
OSCE	Organization for Security and Co-operation in Europe
PFP	Partnership for Peace [NATO]
PGCC	Persian Gulf Cooperation Council
PKK	Partiya Karkerên Kurdistan [Kurdistan Workers' Party]
REFORGER	Return of Forces to Germany
SCC	Standing Consultative Commission
SCEPC	Senior Civil Emergency Planning Committee [NATO]
SCO	Shanghai Cooperation Organisation
SEAL	sea-air-land team
UAE	United Arab Emirates
UN	United Nations
WMD	weapon of mass destruction

CHAPTER ONE

Introduction

As of the time of writing, the United States has begun winding down at least major elements of its engagement in Iraq.[1] However this process finally works out, in particular before the end of 2011, when the U.S.-Iraqi Security Agreement calls for the complete departure of deployed U.S. forces,[2] it is clear that the U.S. relationship with Iraq will change significantly compared with what it has been during the last few years. There will be an *after-Iraq* situation of some shape, character, and timing. But what comes after that? What U.S. strategy toward Iraq and, following the shift in U.S. engagement in Iraq, toward the overall region of the Persian Gulf will best meet U.S. interests and best fit within the tolerances of what the United States is prepared to support? (A map of the Persian Gulf region is presented in Figure 1.1.)

The U.S. standoff with Iran continues, although the Obama administration has begun a process that led, in October 2009, to direct negotiations with the Iranian leadership[3] and involved other permanent members of the United Nations (UN) Security Council, Germany, and Javier Solana, the European Union (EU) High Representative for the Common Foreign and Security Policy (CFSP).[4] The original idea of the American approach was for the United States to reassess progress or the lack thereof in regard to Iran's willingness to meet Western expectations vis-à-vis its nuclear program

---

[1] See Obama, 2009a.

[2] Agreement Between the United States of America and the Republic of Iraq on the Withdrawal of United States Forces from Iraq and the Organization of Their Activities During Their Temporary Presence in Iraq, 2008.

[3] For example, Obama, 2009a: "And going forward, the United States will pursue principled and sustained engagement with all of the nations in the region, and that will include Iran and Syria . . . ." See also Obama, 2009b.

[4] European members involved in the negotiations are the UK, France, Germany (known as the "EU3"), plus Solana, who is to be replaced by the UK's Baroness Catherine Ashton, the newly elected High Representative of the Union for Foreign Affairs and Security Policy. This group has been negotiating with Iran since May 2005. See British American Security Information Council, 2005.

Figure 1.1
Map of the Persian Gulf Region

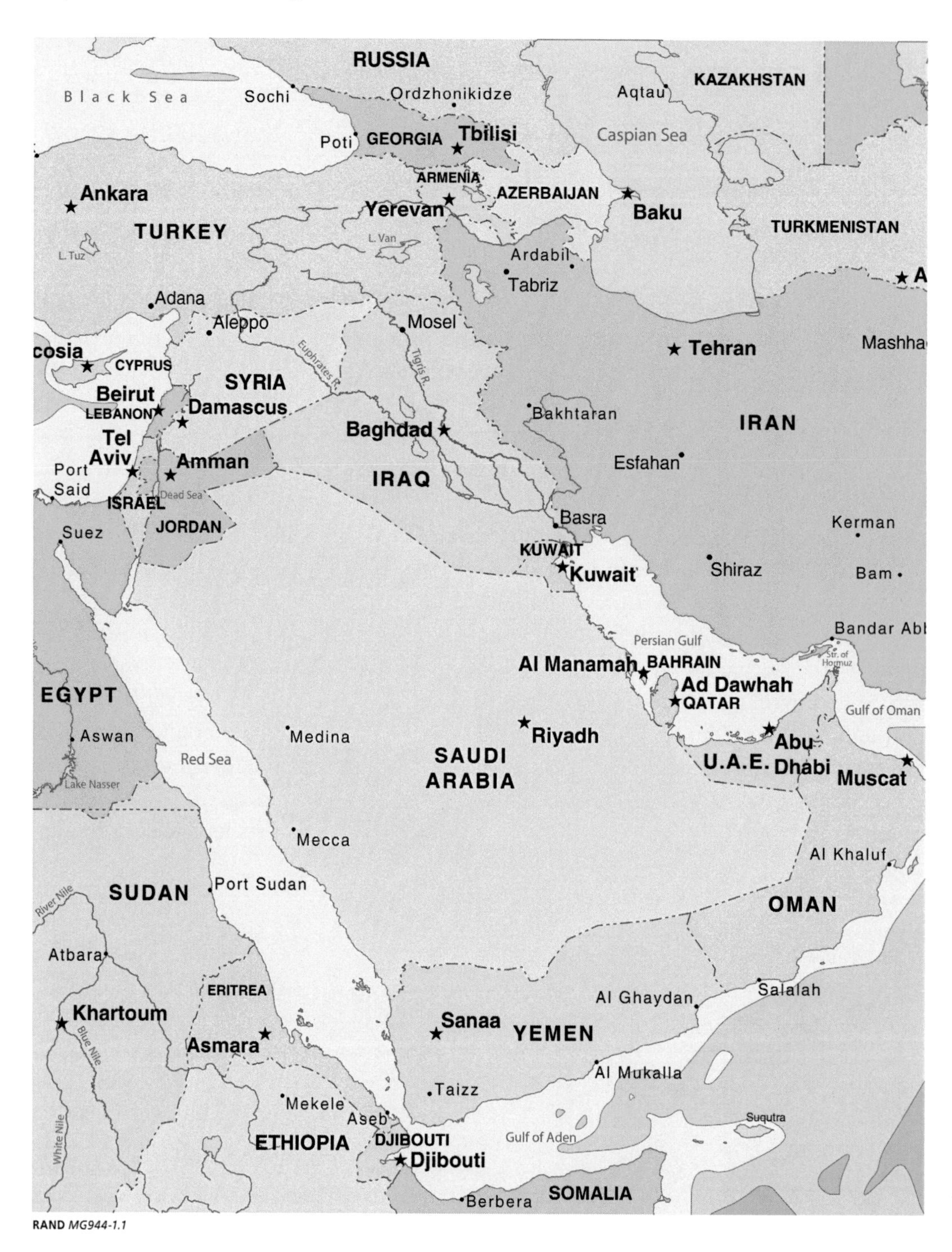

RAND *MG944-1.1*

at about the end of 2009;[5] this goal was later amended to September 2009, although that date (and the end-of-the-year date) passed without major action regarding Iran.[6] What can transpire in the talks that began in October 2009 is another matter. The range of difficult issues between the United States and Iran (and between Iran and some other countries) is extensive, there is a significant legacy of mutual mistrust that will have to be overcome, and the aftermath of the post–June 2009 election turmoil in Iran will have to be factored in. Indeed, this last factor could have a dramatic, perhaps decisive, impact on a number of calculations made in this work. At the same time, there is also a range of issues and areas on which U.S. and Iranian interests are likely to be compatible, though unlikely to be congruent. These also need to be factored into any equation regarding the future of U.S. policy toward Iran.

Yet, concerning the principal concern voiced by outsiders about Iran's policies and actions, what if the country does develop nuclear weapons? The impact of such an occurrence would be highly significant, to say the least, and all the ramifications cannot be clearly understood in advance.[7] One almost certain consequence, within the Persian Gulf region itself, would be the desire of most or even all regional Arab states to gain some form of assurance from the United States against an Iranian military attack or even a lesser Iranian challenge, including efforts to exploit a nuclear capability diplomatically. Thus, considering possible developments, their significance—especially for U.S. interests—and what to do about them will naturally be a major preoccupation of the U.S. government and its allies and friends in the period ahead. And whether or not Iran does develop nuclear weapons or even reaches a point at which it can credibly be argued that it has the capacity to do so, as the saying goes, "with the turn of

---

[5] See the following comment by President Barack Obama on May 18, 2009:

> My expectation would be that if we can begin discussions soon, shortly after the Iranian elections, we should have a fairly good sense by the end of the year as to whether they are moving in the right direction and whether the parties involved are making progress and that there's a good faith effort to resolve differences. That doesn't mean every issue would be resolved by that point, but it does mean that we'll probably be able to gauge and do a reassessment by the end of the year of this approach. (Obama and Netanyahu, 2009)

[6] See the following comment by President Obama on July 10, 2009:

> Now, there is—the other story there was the agreement that we will reevaluate Iran's posture towards negotiating the cessation of a nuclear weapons policy. We'll evaluate that at the G20 [Group of Twenty Finance Ministers and Central Bank Governors] meeting in September. And I think what that does is it provides a time frame. The international community has said, here's a door you can walk through that allows you to lessen tensions and more fully join the international community. If Iran chooses not to walk through that door, then you have on record the G8 [Group of Eight], to begin with, but I think potentially a lot of other countries that are going to say we need to take further steps. (Obama, 2009d).

[7] This point also begs the question whether Iran would be permitted to obtain nuclear weapons or whether Israel, the United States, or both together would take military action to prevent this threshold from being crossed. Officials from the Obama administration have been consistent in saying that no option, including the military option, has been taken off the table. See, for example, the following comment made by Vice President Joseph Biden: "'Still . . . I think it's very important, as we deal with Iran, that we don't take any options, including military options, off the table'" (quoted in Knowlton, 2009).

a screwdriver," will, like postelection developments in Iran, have a major impact on a number of the calculations made in this work. Some of the key factors will be assessed in later chapters, including in a discussion of whether Iran is likely to be cooperative in regard to the development of a regional security structure. Even if a nuclear-power or near–nuclear-power Iran wanted to be cooperative, the political, psychological, and security context could be such that the opposite judgment about Iran could be well-nigh universal on the part of others, at least in the West.

In any event, short of the military destruction of Iran as a viable state (an eventuality with potentially horrendous consequences throughout the region), Iran's continued existence, presence, ambitions as a regional power, and major importance both inside and beyond the region will be facts of life, almost regardless of the nature of its government. It cannot be ignored. Any viable U.S. strategy for the region must therefore account for the Iran factor in some significant form over the long term, even if the nuclear issue were completely resolved to U.S. and Western satisfaction and even with a different government—a point often ignored in analyses that see the nature of Iran's current regime as the only obstacle to an Iran that would do what the United States wants.

Furthermore, the nature of U.S. and Western interests in the region—e.g., the secure export of energy, stability and predictability, counterterrorism, relations with other great powers, geopolitics and geoeconomics in general—means that the United States will have no choice but to remain a deeply engaged power in the region even though the terms, conditions, qualities, dimensions, and application of that power and related influence are subject to debate, decision, and responses to events that have not yet happened or, perhaps, even been imagined. The United States' fate in this regard began to be determined as early as the development of the Truman Doctrine of the late 1940s, and that fate was finally sealed with the invasion of Iraq in 2003.

Further, the choices that the United States makes—alone or with other countries, inside and outside the region—with respect to its own interests, attitudes, and engagements will heavily influence how the United States is viewed, in terms of power, influence, staying power, and *wisdom*, not just throughout the region but also in other parts of the world. The U.S. reputation in terms of these qualities was not significantly damaged by the Vietnam War because the United States' stewardship during the Cold War of the confrontation with the Soviet Union and European communism was essentially well executed.[8] Today, unlike the case during the Vietnam War, which was far removed from the principal theaters of U.S.-Soviet confrontation, the primary locus

---

[8] A key reason that the United States' loss of the Vietnam War did not have a significant impact on the United States' reputation for pursuring its strategic purpose was President Richard Nixon's emulation of the 19th-century British statesman George Canning. In his 1826 expression of approval of the Monroe Doctrine, UK Foreign Minister Canning told the UK Parliament, "'Contemplating Spain, such as our ancestors had known her, I resolved that if France had Spain it should not be Spain with the Indies. I called the New World into existence to redress the balance of the old" (quoted in Bloy, 2002). Nixon cast China in the role of the New World.

of U.S. strategic engagement, in terms of perceived challenges to U.S. security, is not remote from the zone of active conflict. That strategic locus very much includes the Persian Gulf, the broader Middle East, and Southwest Asia, where the major conflicts in which the United States is engaged are taking place. Thus, as a result of the United States' declaration that threats and challenges emanating from this part of the world affect its core security interests, U.S. policies and actions in these three interconnected regions will be judged in a wide context and will have significance for its overall global posture and reputation. During the endgame and beyond in Iraq, it is not just countries in the immediate region—prominently, Turkey and the Gulf Cooperation Council (GCC) countries[9]—that will have questions about the reliability and staying power of the United States, specifically in regard to their own security concerns; countries in the broader Middle East (e.g., Israel, Jordan, and Egypt), in the adjacent region of Southwest Asia (most importantly Afghanistan and Pakistan), and in Europe and elsewhere will have questions as well.[10] Other great powers, notably, Russia, China, and India, will also make judgments about future U.S. strategic policies, commitments, and engagements, and those judgments will be based, in some part, on the choices the United States makes in the fairly near future regarding the Persian Gulf and environs and on what it does to implement and sustain those choices.

---

[9] Bahrain, Kuwait, Oman, Qatar, Saudi Arabia, and the United Arab Emirates (UAE). The GCC countries also cooperate with Egypt and Jordan in the framework of the GCC Plus Two. See, for example, Foreign Ministers of the Gulf Cooperation Council Plus Two, 2007. The U.S. secretary of state also took part in this meeting.

[10] The degree to which such questioning can erupt quickly can be seen in the events after the Russo-Georgian fighting over South Ossetia in August 2008. The fighting followed a declaration by the North Atlantic Treaty Organization (NATO) allies at their April 2008 Bucharest summit that Ukraine and Georgia "will become members of NATO" (Bucharest Summit Declaration, 2008). For each of the NATO allies, although few (if any) understood the full impact of what they were doing, the making of that statement was itself the actual moment of strategic commitment. But when Georgia became embroiled in a sharp but limited conflict with Russia, no ally, including the United States, was prepared to take military action on Georgia's behalf—that is, they defaulted on the implicit commitment they had made to Georgia at Bucharest. A combination of Russian actions and (lack of) Western response immediately led recently joined NATO allies, such as Poland, to question the worth of the Alliance's Article 5 commitment and raised some doubts about the worth of even bilateral U.S. security commitments.

CHAPTER TWO

The Basic Framework

A basic purpose of this work is to determine what, as it reshapes its attitudes, policies, approaches, and engagements regarding the Persian Gulf region, the United States can usefully do, along with others in and out of the region, to develop, foster, and implement a new security structure for the region. A cardinal principle of that security structure should be to provide both benchmarks and a strategic perspective on the basis of which to judge individual policies; to create a set of principles, understandings, and, possibly, some institutions; and to guide U.S. policies and activities and those of others. Ideally, the security structure would allow the United States to be able to promote and achieve its long-term interests—and those of its friends and allies—at a lower cost, both human and material; with less risk of conflict than would otherwise be true; and with lower opportunity costs, thereby increasing the United States' capacity to devote time, attention, and resources to other challenges in the world. As other cases have shown, a security structure that is commonly understood and, in its broadest dimensions, generally accepted by most of the relevant players can yield benefits. It can help to channel relationships, even adversarial relationships, in directions that can reduce the risk of conflict by increasing the capacity to predict the cost-benefit ratios of conflict.[1] It can also help regulate the conduct of conflicts when they occur; and it can reduce costs and

[1] This is always an imperfect measure. Countries and leaderships historically have tended to overestimate their capacity to prevail in conflict and have underestimated the costs in terms of both blood and treasure. *Would we have gone to war if we had understood the consequences in advance?* is one of the most common cries of leaderships and populations after the fact. This is true even when leaders believe that they have learned the lessons of history before embarking on conflict. Thus, the United States thought that its forces in Vietnam at Khe Sanh would not suffer the French fate at Dien Bien Phu in roughly similar battle conditions and, thus, French Emperor Napoleon Bonaparte still invaded Russia even after having first immersed himself in the history of what went wrong when Sweden's King Charles XII had tried the same thing a century earlier. Because of the extraordinary difficulty in making such calculations about the course of conflict (the "fog of war") and because of the tendency of people to believe that they are somehow exempt from the laws of history (a type of self-delusion endemic to the human race), when preventing conflict really matters, the margins of risk have to be widened. Thus, during the Cold War, when a nuclear conflict could have been mutually suicidal, the United States and the Soviet Union worked out attitudes, doctrines, weapons, and procedures that would signal that even relatively small gains could not be pursued without very high costs because of the risks of escalation, however unwanted. The difficulties of managing this process, which could be called *structured overreaction* or *the uncertainty principle*, without actually increasing the risks of catastrophic failure through accident, led both sides to embrace second-strike deterrence.

shape perspectives and expectations regarding the most-prominent and most-potent actors. For example, Cold War confrontation in Europe came to be organized around such a security structure or, more precisely, around the intersection of two contending but complementary and indeed interdependent security structures (one Western, one Eastern). Together, these structures provided a framework for reasonably accurate analysis and prediction about the behavior of one side or the other and, thus, a means for mitigating risk that benefited both sides as well as countries caught, as it were, in the middle.[2] Indeed, it was the success of these two security structures and their interaction, which provided a high degree of predictability, that reduced risks radically and helped, in the end, to bring the Cold War to a peaceful close.

This work is a comprehensive analysis neither of all the security and related factors in play in the Persian Gulf region and beyond nor of all the policy alternatives facing the United States and others, including its allies and partners. That is a vast subject full of complex factors, many of which are in continuous flux and some of which are truly incommensurate, meaning that predictions about local events and the actions of different actors are often very difficult to make—in contrast, say, with the behavior of most European states and nonstate actors, at least in recent times. There will be a high incidence of surprises. It could even be said about the Middle East that the biggest surprise is when there are no surprises:[3] Indeed, despite post hoc analysis and opinion that has argued that the United States should have been able to predict a possible attack on the homeland by Al Qaeda—an argument redolent of post–Pearl Harbor armchair quarterbacking—the 9/11 attacks were, at the very least and even with excel-

[2] This mutually interdependent security structure was completed by the early 1960s after (1) the completion, in the mid-1950s, of the two security blocs and relevant structures, which left no geographic gray zones in the heart of Europe other than, theoretically, Switzerland, Lichtenstein, and Austria and (2) the development of U.S. and Soviet capacities for second-strike deterrence (i.e., assurance that, even during a nuclear attack, the retaliatory weapons of the country under attack could be successfully delivered against the attacking nation and cause unacceptable damage—in this sense, meaning well out of proportion to whatever would be gained by the initial attack). This capacity for second-strike deterrence was enshrined in the extraordinary doctrines and practices that came to be known, at least in the West, as *mutually assured destruction*. These doctrines and practices heightened the predictability of consequences on the part of both sides, provided a high degree of stability despite the horrific consequences for the world if the structure were to collapse, led to détente, and eventually helped lead to the collapse of the Soviet East European empire and the Soviet Union itself.

[3] Then–Secretary of Defense Donald Rumsfeld put the point clearly, succinctly, and accurately, although he was mocked for doing so:

> The message is that there are no "knowns." There are thing [sic] we know that we know. There are known unknowns. That is to say there are things that we now know we don't know. But there are also unknown unknowns. There are things we don't know we don't know. So when we do the best we can and we pull all this information together, and we then say "well that's basically what we see as the situation," that is really only the known knowns and the known unknowns. And each year, we discover a few more of those unknown unknowns. It sounds like a riddle. It isn't a riddle. It is a very serious, important matter. (Rumsfeld, 2002)

lent hindsight, a tactical surprise, although with strategic impact.[4] Even the Soviet invasion of Afghanistan in 1979 and the Iraqi invasion of Kuwait in 1990 caught many U.S. analysts unawares, as did both the pace of the Cold War's end and the extent to which the Soviet Union, its suzerainty in Central and Eastern Europe, and European communism collapsed. Hence the need for a certain degree of humility both in the analysis that follows and in the United States' policy toward the region and the need for an effort to build in latitude for potential surprises without being able to denominate them in advance (otherwise, of course, they would not be surprises).[5]

This analysis will focus on setting parameters for a regional security structure designed to have a high degree of likelihood of covering the key factors in play. Surprises, to the extent possible, are accounted for in the realm of the lesser included.

Accounting for surprises in the structure and operation of any system of security relating to the region of the Persian Gulf and the broader Middle East is made even more difficult because some principal actors are not states and do not conduct themselves according to any (or, at least, to a critical mass) of the norms and expectations of state behavior. Thus, their own behavior cannot be affected—or, at least, cannot be affected to the same degree and in the same manner as state behavior—through the application of rewards or punishments (to put the range of choices narrowly) that can be applied to most states. Not having societies, territory, governments or regimes, and other assets to be put at risk or to be buttressed, some nonstate actors thus fall outside the normal pattern of calculations about cause and effect. Indeed, some of these nonstate actors—especially the practitioners of terrorism—view falling outside that pattern as an asset.[6] This element of unpredictability is different from the added com-

---

[4] Indeed, the 9/11 attacks were probably the most heavily leveraged military act in history in terms of the magnitude of the consequences that have flowed from the deaths of approximately 3,000 people.

[5] While helping to redesign NATO for the post–Cold War world, the author created a concept and an associated statement regarding the future of NATO's relations with Russia, depending on how Russia developed: NATO needed to "bracket" the issue. Thus, if Russia became a successful, democratic, and nonthreatening state, then its engagement in and with Western institutions would enable it to be a full participant and a significant contributor to the historic vision of President George H. W. Bush to create "a Europe whole and free and at peace" (G. H. W. Bush, 1989). But, if Russia went badly from the West's point of view, the United States and NATO would be in a position to oppose Russian behavior through a new containment policy or some other set of policies. Of course, the trick was to pursue the former outcome while preserving a capacity to deal with the possibility of the latter outcome, all without taking action that could undercut the process of fostering a Russia that developed in a positive direction and, thus, permitting development of a long-term positive Russo-Western relationship.

[6] The author uses the term *terrorism* in what he believes is its most correct and most useful sense: an act of violence perpetrated (or threatened to be perpetrated) against noncombatants, on a more-or-less random basis, in an effort to get a wider membership of the class of noncombatants to identify with the victims and thus to put political pressure on a government or other controlling entity to do things that the terrorists want. This is different from nihilism, which is doing violence for its own sake (or, in the American vernacular, "going postal") with no political purpose to advance. Nihilists are different from anarchists: The latter have in mind the destruction of government, not just violence for its own sake. Whenever random violence has a political motive, even if unexpressed or even unintended, it is terrorism.

plications inherent in calculating requirements for a viable security structure caused by the growth of ideology of any sort—itself a phenomenon caused, in major part, because understanding thought patterns based on ideology is alien to most Americans. In recent years, in the Persian Gulf and the broader region, of course, the most-potent such ideology has had a religious or quasireligious cast that is mostly reflected in different aspects of Islam or distortions of Islam. This is especially true of what, for want of a better term, can be summarized as *Islamist extremism*.

In analyzing the elements of a viable security structure for the region, it is necessary first to differentiate a security *structure* from a security *system*. It could be argued that, in any region that is not in total turmoil (and, perhaps, even then), there is always, perforce, a security system. This system consists of some vision or perception of relationships between different countries and peoples that includes assessments, commonly shared or not, of what those relationships are and how they could be affected by various actions taken or postures struck.

As the terms are used in this work, a security *structure* is more than just a security *system*. A structure is some organization of relationships among countries (and non-state actors) that supplies at least a reasonable level of predictability about what different actors will do within a reasonably likely range of circumstances plus feedback that informs different parties to the overall structure regarding the impact of actions taken. There is clearly an emphasis on something formal as opposed to simply happenstance, and there is also an implication that arrangements have some institutional element, as opposed to being simply ad hoc.

From the perspective of any particular participant—in this case, the United States—a security structure is most effective and useful when it not only enables the participant to pursue its interests at any point of time with at least a reasonable prospect of success but also offers a reasonable prospect that the same will be true in the future. A valid and useful security structure, therefore, is not a one-off but rather something that has some capacity to endure and, hence, offer predictability. *Thus, for the United States, an ideal security structure for the Persian Gulf is not just one that will help the United States, to the degree possible, to secure its interests today but also to do so tomorrow; indeed, such a structure will be able to accommodate significant shocks from either within or without.* Again, predictability is a key objective; it is also a key asset.

Another word about process is needed here. This work does not concentrate on a detailed schema for dealing with every immediate issue, such as the best way for the United States to disengage from Iraq or how to structure and conduct particular negotiations with different countries. Instead, it takes as its mandate the devising of a set of criteria for a regional security structure that could be more profitable, for the United States and others, than the one that is otherwise likely to emerge from current circumstances and from the likely course of both events and policies in regard to those events. That is, in the analysis that follows, there is a quality of reaching for the ideal: What security structure would be more mutually profitable in the future and, therefore, increase predictability; reduce the risk of conflict (if not the material costs of providing security); and *reduce the United States' exposure, its costs, and, ideally, its responsibilities in the region over time.* In short, what is presented and analyzed in this work is, in the first instance, a series of benchmarks—something to be worked toward because of its inherent value for the United States and other participants in the structure. To borrow from language used during the U.S.-led efforts after the Cold War to create "a Europe whole, free, and at peace,"[7] this would be a plus-sum, rather than a zero-sum, game. Positing an ideal, however, does not mean just reaching for a pie in the sky. Instead, it is an effort to set criteria, and to analyze courses of action based on those criteria, that can set up possibilities that can be strived for. It is not simply formulating a wish list that only meets Robert Browning's standard of "a man's reach should exceed his grasp." Security in the Persian Gulf region cannot wait for heaven but must be sought in the relatively near-term future here on earth.

---

[7] This was President George H. W. Bush's grand strategy for the Continent, which set the criteria for all actions taken since by the West to restructure European security: "It is time to offer our hand to the emerging democracies of Eastern Europe so that the continent, for too long a continent divided, can see a future whole and free" (G. H. W. Bush, 1990).

CHAPTER THREE

Background and Context

The Lessons of 1979 and Beyond

For the purposes of this analysis of terms and conditions for creating a new and viable security structure for the region of the Persian Gulf, the benchmark moment for comparison of the past and the future is chosen, not all that arbitrarily, as 1979.[1] To be sure, the Persian Gulf and the broader Middle East had not been all that stable even before that time, although, except during the periods of direct combat between Israel and its neighbors (in 1967 and 1973), there had not been significant shocks to the system of regional security since about the 1950s, the time of the early Arab Revolution and its manifestations in Egypt, Syria, and Iraq. (Crisis and conflict also occurred, from time to time, in Lebanon.)

From the standpoint of regional stability and a reduction of the risk of conflict, 1979 brought about one positive development: the conclusion of the Egyptian-Israeli Peace Treaty, which effectively reduced almost to nil the possibility of open conflict between these two countries and, therefore, also lowered dramatically the likelihood of any military assault on Israel by any Arab army. Even when Egypt was part of the Arab-Israeli military balance, Israel had prevailed in three major conflicts. With Egypt out of the balance after the treaty, an attack on Israel by any Arab state or combination of Arab states became an act of folly, at least until the recent development of the modern form of asymmetric-warfare techniques. This strategic shift had an added effect: It reduced virtually to zero the risk—which had existed in small measure during the 1967 Six-Day War and in much greater measure during the 1973 Yom Kippur

[1] That year was not, of course, the first time the United States had become deeply involved in the affairs of the Persian Gulf or even had become involved in efforts to create some form of security structure for the region. Perhaps the most notable of such efforts was the Baghdad Pact of 1955 (involving Turkey, Iraq, the UK, and Pakistan), which—after Iraq's withdrawal—was replaced in 1959 by the Central Treaty Organization (CENTO), whose regional members were Turkey, Pakistan, and Iran. Although the United States did not join CENTO, it reached bilateral agreements with the three regional states. After the Islamic Revolution, Iran withdrew from the organization, and CENTO was disbanded in 1979 (see U.S. Department of State, undated). CENTO never had a military command and was most useful in terms of economic and technical assistance. Primarily because it was a Cold War entity, had a limited membership, and lacked significant security cooperation among its regional members, CENTO is not cited in this work as a possible model for a Persian Gulf security structure.

War—of a U.S.-Soviet confrontation over the Arab-Israeli conflict. For the United States, therefore, the Egyptian-Israeli Peace Treaty not only increased the chances of stability and peace in at least one part of the Middle East, with its attendant spillover effect in other parts, including the Arab states of the Persian Gulf: It also ratified the United States' strategic primacy throughout the region, which was a major geopolitical development.[2]

However, 1979 also brought two assaults on the United States' understanding of, assessments of, and confidence about the existing security system in the Persian Gulf and nearby areas. One of these assaults was Iran's transformation, through its Islamic Revolution, from a country supportive of U.S. security objectives in the region—a so-called regional influential[3]—into a country definitely at odds with the United States both strategically and ideologically. This transformation was seen as having a potentially significant impact on U.S. relationships with other countries and societies (at least Shi'a-denominated societies) in the immediate region and beyond. Indeed, the fallout from the Islamic Revolution, including the 444 days during which U.S. employees of the U.S. embassy in Tehran were held hostage, has since affected U.S. perceptions not just of its relations (or nonrelations) with Iran and threats and other challenges emanating from Iran but also of the entire compass of security within the region, extending at least as far as Afghanistan and Pakistan to the east and Israel and Lebanon to the west, possibly including Libya and Algeria.[4]

The Islamic Revolution—which was followed by a hostage crisis that directly affected the United States not as a strategic act but as a very much human issue and a humiliation—was also the forerunner of a new form of challenge to the United States, at least to a degree and of a kind that it had rarely faced before.[5] It was the United States' introduction to asymmetric warfare practiced not as part of a broader military conflict, a war of "national liberation," or an effort to achieve territorial gains but rather for ideological purposes. In this case, the purpose was to gain advantage through humiliating

---

[2] This also completed the transition from the era of British and French colonialism and the two countries' roles as competitors for power and influence in the region to the United States' assumption of its exclusive position as the dominant and unchallenged Western power in the region. (The treaty also greatly diminished the Soviet Union's prospects of becoming a major power in the region—a reduction for which the Soviet Union's invasion of Afghanistan, even had it been successful, could not compensate.)

[3] A term coined in 1977 by then–U.S. National Security Advisor Zbigniew Brzezinski. See Brzezinski, 1983, pp. 53–56.

[4] For an excellent description and analysis of this period in U.S.-Iranian relations, see Sick, 1986.

[5] The United States experienced this form of asymmetric warfare in Vietnam, where one of the U.S. enemies' goals was not just to affect circumstances on the ground but also to affect domestic public and political opinion in the United States. Asymmetric warfare of a more classic sort, now sometimes subsumed under the term *insurgency*, has been evident in many guises over many years, including as a staple of the limited-war aspects of the Cold War. The United States faced these techniques during the Philippine-American War in 1899–1902 (with fighting continuing until 1913); in fact, significant elements of U.S. tactics during the Revolutionary War fit definitions of *insurgency*. See Fischer, 2004.

the Western superpower and to demonstrate to other countries, in the region and elsewhere in the Islamic and Third Worlds, that U.S. power could be countered—in this case, through a technique that led the United States to be effectively self-deterred in its responses because available responses, including the use of military force, were judged as likely to be counterproductive.[6] In addition, this countering of U.S. power produced internal political benefits for the Iranian regime.[7]

Since that time, the technique of asymmetric warfare, including its ideological element, has been used more frequently against the United States, most notably in the 9/11 attacks and, more recently, during the ensuing conflicts in Iraq, Afghanistan, and Pakistan. Since 1979, the United States and the West in general have faced the problem of ideological contagion among some elements of the Islamic world, both Shi'a and Sunni. Thus, in November 1979, during the Iranian hostage crisis, when Saudi Arabia's Grand Mosque in Mecca, Al-Masjid al-Haram, was seized by Islamic fundamentalist dissidents and erroneous reports of a U.S. attack on that mosque surfaced, the U.S. embassy in Islamabad, Pakistan, was besieged and burned, nearly resulting in a great number of deaths.

The other principal shock of 1979 that had a major impact on U.S. strategic calculations about the region of the Middle East and beyond and, thus, on the United States' sense of a strategic framework for the broader region was the Soviet invasion of Afghanistan. This reshaped U.S. thinking about what the Soviet Union was prepared to do within the context of informal yet robust understandings with the United States on the limits of each state's conduct in the Cold War. Although it could be argued that Afghanistan was outside the geographic center of U.S.-Soviet confrontation—that is, outside the most-neuralgic areas of competition between the two nations and ideologies—the United States chose not to see it that way. This was, in part, because Afghanistan was close to the oil-exporting region of the Persian Gulf; in part, because of concerns that the Soviet Union might exploit the U.S.-Iranian standoff during the hostage crisis, along with Iran's internal political turmoil, to move militarily into Iran;

---

[6] Until the failed U.S. hostage rescue mission of April 1980, the United States was effectively deterred by two factors: (1) concern that military action taken to free the hostages or to punish the Iranian regime (i.e., compellence) would get many or all the hostages killed and (2) concern that a punitive U.S. military attack would simply play into the hands of Ayatollah Ruhollah Mousavi Khomeini, especially if it made him a martyr. The rescue mission was mounted, in part, because of growing evidence that the lives of the U.S. hostages were at risk. The author, who was, at the time, a member of the National Security Council staff and held senior responsibilities for the Middle East, believes that the hostage rescue mission, although ostensibly a failure, very likely saved hostage lives by demonstrating to the Iranian regime that there was a point beyond which the United States could not be pushed without taking direct and significant military action.

[7] While serving on the National Security Council staff at the time, the author judged that Ayatollah Khomeini had not precipitated the seizure of the U.S. embassy but that he had quickly started exploiting it to consolidate his domestic political base. Indeed, every movement in the direction of resolution of the crisis came after Ayatollah Khomeini had achieved a domestic political goal (e.g., the adoption of the constitution he wanted and the election of officials he wanted).

and, in part, because of a continuing sense that aggressive actions by the Soviet Union in one place that went unchallenged could have a seriously negative impact elsewhere, including in others' perceptions of U.S. reliability in countering challenges by the Soviet Union that involved direct military invasion of another country.[8]

These twin crises that called into question critical U.S. assumptions about security in and around the Persian Gulf and the broader region of the Middle East and Southwest Asia led the United States to launch a number of initiatives. Two stand out for the purposes of this work. One was the effort to impose costs, of one sort or another, on the Soviet Union over the issue of Afghanistan; this effort included the start of active, covert U.S. engagement with the Afghan resistance.[9] At the same time, the Carter administration was concerned that the Islamic Revolution and the hostage crisis could lead the Soviet Union to be emboldened to strike at Iran in addition to Afghanistan. From the point of view of attentive international audiences, the United States either had chosen not to do very much about Soviet actions in Afghanistan or was unable to do so. Would the United States also prove impotent, by accident or by design, in regard to possible Soviet ambitions in Iran? Thus, the administration propounded the Carter Doctrine, which President Carter announced in his 1980 State of the Union address:

> Let our position be absolutely clear: An attempt by any outside force to gain control of the Persian Gulf region will be regarded as an assault on the vital interests

---

8 The Vietnam War was, in part, a conflict by proxy between the Soviet Union and the United States that fell outside of the commonly understood zones of significance to either of the parties and was somewhat analogous to the way in which Britain and Russia played out their mid-19th–century competition by fighting in the Crimea rather than in some other place where either or both could have gotten into more trouble than the conflict would have been worth. The Korean War was also a conflict beyond the critical areas. The United States decided to intervene, and the Soviet Union did not, at least not in any acknowledged way, although it was assumed from the beginning that Moscow had instigated the war, and it has become clear that Soviet military personnel were involved in air combat.

Conflict by proxy is a relatively common historical phenomenon in great-power competition. In 1979–1980, the United States was concerned that the Soviet Union could misinterpret the West's relative nonresponse to the invasion of Afghanistan as license to try advancing against Western interests elsewhere in the immediate region, although the United States did not believe that the invasion presaged Soviet attacks against critical Western assets in Europe or in the Far East (e.g., in Japan or South Korea). The United States did seek to send a message to the Soviet Union about the limits of Soviet actions that crept close to the boundaries of the two countries' implicitly agreed "do-not-go" understandings. Parts of this message were substantive, such as U.S. support for Afghan insurgents against the Soviet occupiers. Some—such as the U.S.-promoted boycott of the Moscow 1980 Olympic summer games, which seems to have had no appreciable impact on the Soviets but did further sour perceptions about the administration of President Jimmy Carter and help lead to his defeat in the elections that fall—were symbolic. The author had counseled that the United States participate in the Olympics and "win one for the Gipper" in regard to Afghanistan.

9 In one of history's great ironies, the United States helped to train resistance fighters for the struggle against the Soviet Union in Afghanistan, only to see some of them reemerge in the ranks of Al Qaeda.

of the United States of America, and such an assault will be repelled by any means necessary, including military force.[10]

This declaration was, in fact, implicitly directed against a possible Soviet military or other aggressive move against Iran, a country then locked in confrontation with the United States. A search through history may not come up with any similar, seemingly bizarre strategic commitment to a country that was, at the time, an enemy.[11] This U.S. posture has a further instructive element: If, in 1980, it was possible for the United States to look beyond its current confrontation with Iran because of a more-important strategic interest (i.e., containing the Soviet Union), perhaps the United States can, at least to the degree necessary, look beyond today's confrontation with Iran's Islamic government in order to promote longer-term U.S. interests in the region, not so much in regard to the Iranian nuclear program as to the future of Iraq and Afghanistan, the effort to defeat terrorism, and—following the logic of this work—laying groundwork for a future security structure in the region that would permit the United States to achieve its objectives at less risk and cost than at present.

The United States also recognized the need to reassure other states in the Middle East, especially in the Persian Gulf region, both against the possibility of a further Soviet thrust (a small risk) and against the possibility of direct destabilization or worse practiced by the revolutionary Iranian regime (a large risk). This reassurance included the need to preserve the United States' reputation for doing something effective in terms of preserving and advancing its political and security interests, a reputation placed in jeopardy by the hostage crisis (although there was no direct threat to the U.S. homeland).[12] Regional states, along with the rest of the world, could see that, in its efforts to try to free the hostages, the United States was mounting the most-massive

---

[10] Carter, 1980. The author led a small interagency working group that fashioned the parameters of the Carter Doctrine, and he drafted much of the language in the State of the Union address. One fortunate turn of events, as it were, was that Iraq only invaded Iran in September 1980. Had it done so a year earlier, at the time of the Soviet invasion of Afghanistan, the Soviets might have been tempted to extend their military venture into Iran. At least, fears thereof would have been heightened even further in the United States.

[11] A partial parallel can be found in fears that a nuclear-armed state might end with a bang, not a whimper and thus take others down with it in a nuclear spasm. This was one of the reasons that the United States and other Western states were reluctant during most of the Cold War to pursue active policies to try bringing down the Soviet regime—begging the question, of course, whether that could be achieved. In terms of strategic commitment to the region by a Great Power, the Carter Doctrine bore a remarkable resemblance to the classic statement of British policy by Lord Lansdowne in 1903, when he said that the United Kingdom would "regard the establishment of a naval base, or a fortified port, in the Persian Gulf by any other power as a very grave menace to British interests, and we should certainly resist it with all the means at our disposal" (quoted in Sick, 2003, p. 294).

[12] It can be argued that U.S. policy during this crisis, including its permitting itself to be deterred from taking direct military action (other than the abortive hostage-rescue mission that occurred midway through the crisis and that was not a *punitive* exercise) increased the chance that the United States would, in the future, be exposed to asymmetric warfare of one form or another (e.g., efforts to affect U.S. policy through relatively small actions with a potentially big impact on U.S. domestic public opinion).

diplomatic undertaking in its entire history. Therefore, these countries could be forgiven for drawing negative inferences about the United States after the hostage-rescue mission failed in April 1980.[13]

Thus, the United States undertook a number of efforts at reassurance. These included air force exercises with the Egyptians; securing forward operating bases in Egypt (at Ras Banas) and Somalia (at Berbera);[14] gaining from Britain more-extensive rights to operate militarily from the Indian Ocean island of Diego Garcia; beefing up the U.S. Naval Support Activity in Bahrain (in use by the U.S. Navy since 1948); reaching agreement with the Sultanate of Oman to make (limited) use of military facilities;[15] and creating a Rapid Deployment Joint Task Force, the forerunner of today's U.S. Central Command. In sum, these moves were fashioned around the idea of an informal security structure for the region that was based, however, not on a coalescence of actions and understandings by and with regional states but on the projection of U.S. military power.[16]

At the same time, these military moves were designed less to give the United States the capability of engaging in combat and more to reassure local friends and allies of an implied but not openly stated U.S. commitment. In this case, aggression from the Soviet Union was not the issue; rather, the issue was aggression from Iran or other potential mischiefmakers inspired by the Islamic Revolution.[17] Indeed, one critical element of these U.S. activities was to develop the U.S. capacity for operating militarily over the horizon.

---

[13] It could even be argued that the intensity of the United States' efforts and its clear preoccupation with the hostage crisis actually made matters worse in terms of its reputation for decisive action: The bar was raised, but the United States did not clear it. This may help to explain why President Ronald Reagan, by contrast, played down the significance of the seizure of a small group of American hostages in Lebanon in 1981. President Carter had followed a "Rose Garden strategy" of keeping close to the White House during most of the Iranian hostage crisis, a strategy that served to heighten expectations about his capacity to resolve the crisis. President Reagan learned from President Carter's experience, seeing the political costs of this decision, and did not follow suit.

[14] On Ras Banas, see Lefebvre, 1991, p. 232. On Berbera, see H. Cohen, 2002: "With Berbera becoming a key component of U.S. military planning in the defence of the Persian Gulf region, U.S.-Somali relations became even more important to Washington." Military aspects of this relationship came to an end in 1989 at the insistence of the U.S. Congress.

[15] According to U.S. Department of State, 2007,

> U.S.-Omani relations were deepened in 1980 by the conclusion of two important agreements. One provided access to Omani military facilities by U.S. forces under agreed-upon conditions. The other agreement established a Joint Commission for Economic and Technical Cooperation, located in Muscat, to provide U.S. economic assistance to Oman.

[16] The deterrent value of these actions needs to be seen in the context of the Soviet Union, not Iran, since little of this added capability would have been needed for punitive strikes against Iran.

[17] Over the years, the United States has also developed close military-to-military relationships with a number of Gulf Arab states, including supplying them with weapons and using their bases.

There was acute sensitivity in Washington regarding the potential reaction among local populations, or even governments, if there were too much U.S. military presence on the ground; this presence could serve as fodder for propaganda by opponents of the United States, who would seize on the political irritant of there being outsiders, especially Western soldiers ("infidels"), both visible and in sizable numbers in Muslim countries.

This point is emphasized here because it is instructive regarding a possible security structure for the present and future Persian Gulf region. *To the extent that the United States will need to be able to deploy forces in the region to advance its interests in regional security, can it station them in places where they are largely out of sight but not out of mind?* In 1979–1980, this was a major concern in U.S. policy and regional activities.[18] Unfortunately, the principle involved was not heeded after the 1991 Persian Gulf War, and significant numbers of U.S. military personnel remained in bases in Saudi Arabia, where they became lightning rods for Islamist opposition.[19] After the attacks of 9/11 and prior to the 2003 invasion of Iraq, Qatar—a country less subject to religious fervor than Saudi Arabia and also concerned about the possible ambitions and intentions both of that large Arab neighbor and of Iran—became the primary locus for U.S. forces deployed on the ground in the region.

Containment and Dual Containment

After these two great shocks during 1979 to the United States' relative sense of security regarding the region of the Persian Gulf came a third in September 1980, when Iraq, taking advantage of perceived Iranian weakness, began what came to be a major eight-year war. This event produced little joy in the United States, where some feared, with some justification, Iranian suspicions of U.S. complicity, added threats to the Americans whom Iran held hostage, and the Soviet Union's exploitation of the situation by deciding to move against Iran even in the face of the United States' commitment under the Carter Doctrine.[20] A minority viewpoint was that the war might hasten release of the hostages, given that Iran was facing a threat far graver than any ideological or other dispute it had with the United States. Neither scenario came to pass.

With the end of the hostage crisis in January 1981, evidence that the Soviet Union was not going to move militarily beyond Afghanistan, and Iran's being under assault

---

[18] While the director for Middle East Affairs at the National Security Council, the author was instrumental in preventing a similar problem regarding the presence of U.S. forces in Egypt.

[19] Osama bin Laden and other leaders of Al Qaeda seized on the U.S. military presence to foster opposition to the United States.

[20] Soviet restraint toward Iran may have had something to do with the Carter Doctrine; if so, the timing of its enunciation, eight months before Iraq's invasion of Iran, was fortuitous.

by Iraq, the United States perceived a situation of relatively greater security in the sense that its own immediate interests did not seem to be at the serious risk that had been feared a few months earlier. Its relationship with Iran was then transmuted into a pattern of containment composed of active efforts in terms of state-to-state relations and watchful concern in terms of Iran's efforts to export its revolution. These Iranian efforts eventually dwindled as potential targets of Iran's propaganda and proselytizing recognized that the revolution was producing little in the way of benefits for the Iranian people and was, thus, not to be emulated. U.S. containment of Iran, used as a means of seeking to promote overall security, stability, and predictability in the region, included substantial support, mostly covert and undeclared, for Iraq in its war with Iran.[21]

U.S. involvement on Iraq's behalf was also a modified balance-of-power strategy, as the United States took Iraq's part against Iran during the Iraq-Iran War, although never so decisively as to ensure Iraq's success. In the midst of the conflict, the United States engaged in a bizarre, almost comic-opera venture of seeking to sell arms to Iran in order to gain funds that could then be used, covertly, to support the so-called anti–Sandinista Liberation Front contras in Nicaragua—a move taken in violation of the explicit wishes of the U.S. Congress.[22] But this was not a calculated balance-of-power strategy; indeed, the type of nonideological, almost mechanical behavior required in a true balance-of-power approach (including switching allegiances as circumstances require to retain balance), as traditionally practiced by Britain, is not particularly congenial to the American way of thinking. Indeed, the United States has had a penchant to identify strongly with the particular side of a competition, struggle, or confrontation it has chosen to support, often for some ideological or imputed ideological reason (e.g., prodemocracy, anticommunism, or, as in today's Middle East, anti-extremism), except when, for either real or imputed strategic reasons, it chooses to dispense with or compromise on these criteria, as it did in supporting both the Soviet Union during World War II[23] and various "friendly" dictators around the world during the Cold War (and still, in some places, today).

---

[21] This U.S. support included provision of intelligence, some weaponry, and, from March 1987 onward, the reflagging and naval escort of 11 Kuwaiti oil tankers transiting the Persian Gulf. Indeed, the accidental shooting down of an Iranian civilian airline in July 1988 by the USS *Vincennes*, which killed all 290 passengers, is credited by some observers as having pushed Iran into both the direction of accepting a ceasefire that began the following month and, in effect, a losing position in the Iran-Iraq War. Iran then accepted UN Security Council Resolution 598 of July 20, 1987. See United Nations Security Council, Resolution 619.

[22] Although President Reagan was severely criticized for the Iran-contra scandal, the idea of helping Iran to resist losing the war to Iraq was strategically sound (although it was doubtful that strategic calculations were involved in that decision). Keeping Iran from collapsing and leaving Iraq as the dominant power in the region was the functional equivalent of the Carter Doctrine vis-à-vis the Soviet Union and constitutes yet another irony of U.S. policy.

[23] The point is introduced here because of the consequences that flowed from the United States' assumption that "the enemy of my enemy is my friend," a belief that helped lead to the United States' delusion that Iraqi President Saddam Hussein would promote U.S. interests as well as his own. The Iraqi invasion of Kuwait in 1990 was, in

The relative stability in the region that was conferred by the end of the Iran-Iraq War in 1988 did not last long. It was followed in August 1990 by Iraq's invasion of Kuwait.[24] The extent to which President Saddam thought he had a "yellow light" from the United States is still the subject of debate; so, too, is the extent to which the United States was lulled into misjudgments about President Saddam because of its recent experience as a de facto ally in Iraq's war with Iran and because of the relationship of that war to the U.S. containment policy toward Iran. In any event, President Saddam was rapidly disabused of any such notion and was expelled from Kuwait during the U.S.-led coalition military operation, Desert Storm, in January–February 1991. Very much relevant to this work is the fact that Iran supported the U.S.-led efforts to expel Iraq from Kuwait, although its motive in doing so was obviously self-interest. Iranian support included the impounding of much of the Iraqi Air Force after it was, bizarrely, flown to Iranian air bases.[25] After the ceasefire, however, the United States stood by while President Saddam carried out military attacks on various Shi'ite groups—including the Marsh Arabs, who lived in an area adjacent to Iran—that were presumably calculated to prevent Iran from gaining a political toehold in Iraq.[26]

Nevertheless, Iran's supportive role in the Persian Gulf War did not lead to a revision of Washington's policy toward Tehran. Instead of reconsidering its containment of Iran, the United States added Iraq and, more particularly, the Iraqi regime of President Saddam to its list of countries and regimes with which it would not deal and that should be contained. This move was organized through an elaborate set of arrangements—political, economic, and military—that were sanctioned by the UN.[27]

---

part, a product of this self-delusion, which kept the United States from warning Iraq of what Washington in fact would do after President Saddam misread what would become U.S. policy after the United States awoke to its own folly.

[24] There is no public evidence that the United States calibrated its involvement with Iraq in order to enable Iraq to do so much and not more to Iran. In other words, there is no evidence that the United States adopted a dynamic balance-of-power approach. Of course, the United States never did provide Iraq with sufficient support to make possible the defeat of Iran, if any practical level of U.S. support could indeed have enabled Iraq to achieve such an outcome. It is also far from clear that, if Iraq were about to occupy a significant part of Iran, the United States would have intervened on Iran's behalf, an action that would have been an essential element of a true balance-of-power strategy. However, U.S. actions did risk being inconsistent with the Carter Doctrine, since Iraq's assaults on Iran could have weakened the latter to the point at which it would have become an easier mark for Moscow. In 1985–1986, during the Iran-Iraq War, the United States became engaged in the covert supply of weapons to Iran. Whether that decision derived from a strategic calculation to keep Iran from succumbing to Iraq, as opposed to from a desire to find a way to support the contras without the knowledge of the U.S. Congress, is doubtful. If the arms supplied to Iran did help keep it in the war as opposed to suffering defeat—thus preventing the domination of the Persian Gulf by Iraq—it could be termed a fortunate, accidental, U.S. balance-of-power policy.

[25] See Gordon, 1991.

[26] See, for instance, Rosenberg, undated. Also see United Nations Security Council, Resolution 688.

[27] See United Nations Security Council, Resolutions 686, 687, 688, 689, 692, 700, 705, 706, 707, 712, and 715.

Single containment thus mutated into dual containment. It has been argued that the inability or unwillingness of the United States to pursue a true balance-of-power policy led it into a situation in which it had to provide the balance of power against both Iraq and Iran,[28] in the sense of the term that came into fashion during the Cold War, when the United States and its allies provided a unitary balance against the Soviet Union, its allies, and its acolytes. The strategy of dual containment, which did not admit of an approach that would favor one side or the other, depending on circumstances and desiring to use the strength of one party to offset the strength of the other, also required that the United States remain deeply engaged in the Persian Gulf to balance or contain both countries at the same time. It could not act as a more-or-less invisible hand, maneuvering the two parties to contain one another at (possibly) reduced cost to the United States. The United States' dual-containment policy included the aforementioned error of keeping U.S. forces in Saudi Arabia after the end of the Persian Gulf War and abstaining from any effort to work with either Iran or Iraq against the other. With both regimes being regarded by the United States as pariahs, it was unwilling to treat with either and thus left itself with no flexibility to try pursuing state interests as opposed to interests heavily colored by ideology and even by a feeling of being betrayed, in different ways, by both regional countries.[29]

What ensued for more than a decade was the U.S. policy formally known as *dual containment*, which kept an uneasy peace and provided the key elements of a form of security system for the Persian Gulf region, although without mechanisms to move this system of security into a form—i.e., a structure—in which it might have acquired some lasting stability. Indeed, the situation was inherently unstable, kept relatively calm only by a continuing U.S. military presence in the region[30] and by the absence of any serious regional attempt to challenge the United States' role. At the same time, gone was any sense that the United States could choose to support, up to a point, either Iraq or Iran to prevent the domination of the Persian Gulf region by the other country. Instead, from 1991 until 2003, the United States chose, for geopolitical reasons, to keep both countries from having significant influence within the Persian Gulf region or beyond: In Iran's case, the United States' motives also included a desire to prevent the export of the Islamic Revolution, although, by then, the revolution's appeal had dwindled significantly. U.S. containment of Iran was essentially passive—a mixture of nonintercourse, economic sanctions, and efforts to limit Iranian relations with third countries. U.S. containment of Iraq, by contrast, was more active, including the

---

[28] The author argued this point in mid-1992, including in a memorandum prepared for Democratic Party presidential nominee Bill Clinton. The memorandum was published in early 1993 by the Clinton Transition Team.

[29] Of course, it is impossible to determine whether U.S. efforts to play Iraq and Iran against one another could have succeeded.

[30] Saudi, British, and French aircraft also took part in Operation Southern Watch. Britain and Turkey joined the United States in Operation Northern Watch. Both operations were mandated by the UN Security Council. See GlobalSecurity.org, 2005a, 2005b.

UN-sanctioned imposition of no-fly zones over the northern and southern parts of the country (Operations Northern Watch and Southern Watch)[31] and various forms of sanctions that were supposedly tailored to minimize the relative suffering of the Iraqi people.[32] The United States did, from time to time, attack targets in Iraq when Iraq violated the no-fly zones, and, with Britain, it attacked "Iraq's nuclear, chemical and biological weapons programs and its military capacity to threaten its neighbors."[33] Indeed, in the Iraqi Liberation Act (signed into law in October 1988), the United States went on record as declaring "that it should be the policy of the United States to seek to remove the Saddam Hussein regime from power in Iraq and to replace it with a democratic government," a policy also endorsed by President Bill Clinton.[34]

The Invasion of Iraq

The aforementioned policy is particularly important to note here because it indicates that the U.S.-led invasion of Iraq in March 2003 was a direct descendent of the U.S. policy of dual containment adopted informally by the administration of President George H. W. Bush, formalized in the administration of President Clinton, and indeed acted upon militarily—i.e., in a move beyond simple containment—on a few occasions by the latter administration. The 2003 U.S.-led Coalition invasion of Iraq was the result of a policy that was different in degree but not in kind from that of the Clinton administration: Under President George W. Bush, containment went from passive-aggressive to active-aggressive. This transition in U.S. policy, applying at least in the first instance to Iraq and not to Iran,[35] was certainly influenced by the shock of the 9/11 attacks on the United States but, as much commentary and analysis have pointed out, there has been no convincing evidence that Iraq (or President Saddam) was in

---

[31] See GlobalSecurity.org, 2005a, 2005b.

[32] See Rempel, undated; United Nations Security Council, Resolution 986.

[33] See W. Clinton, 1998b. By contrast, the United States at no point used military force against Iran, even on occasions when some observers believed that terrorist attacks on U.S. personnel (e.g., the attack on the U.S. barracks at Khobar Towers in Saudi Arabia in June 1996) had been instigated by Iranians. See Eggen, 2004; W. Cohen, 1997.

[34] In signing P.L. 105-338, President Clinton said, among other things, that

> while the United States continues to look to the Security Council's efforts to keep the current regime's behavior in check, we look forward to new leadership in Iraq that has the support of the Iraqi people. The United States is providing support to opposition groups from all sectors of the Iraqi community that could lead to a popularly supported government. (W. Clinton, 1998a)

[35] Despite extensive speculation on both sides of the argument, it is impossible to know whether success in the aftermath of the invasion of Iraq in 2003 would have led the United States to attack Iran as well. Such an attack would have been consistent with the worldview of a number of active and influential supporters of the invasion of Iraq, both in and out of the U.S. government, but whether their counsel would have prevailed with President George W. Bush is an unknowable proposition.

any way connected with that attack.[36] Debate continues over the extent to which this invasion was a product of what has come to be known as the Wolfowitz Doctrine, a policy according to which the United States should use the unmatched military and economic power with which it emerged from the Cold War, along with the absence of a countervailing power (i.e., the Soviet Union), to oppose the development of aspirant regional hegemons with interests and attitudes antithetical to, or at least inconsistent with, those of the United States.[37]

After the formal end on May 1, 2003, of what proved to be only the first phase of combat in Operation Iraqi Freedom, the United States found itself unable to withdraw or retreat from major, deep, direct military engagement in Iraq. As can be gleaned from earlier sections of this work, even before the March 2003 invasion of Iraq, the structure of security in the Persian Gulf region—to the extent that any structure existed at all—was shaky at best, although there was a high degree of predictability about what different parties would do under a variety of circumstances.[38] In short, the old system of security, such as it was, was shattered by the 2003 war, and, since then, the United States, because of its own self-interest, has had no choice but to help shape and develop the system that takes its place.

It is important to understand that a similar injunction applies to the United States' European allies[39] and, indeed, to every other country in the world with an interest in anything approximating stability in the region of the Persian Gulf, partly because of the outside world's dependence on the flow of hydrocarbons from the region.[40] Of course, this statement begs an important question that has been often raised but never

---

[36] See, for example, the following statement by President George W. Bush: "It is true, as I've said many times, that Saddam Hussein was not connected to the 9/11 attacks" (G. W. Bush, 2008). See also "Cheney: No Link Between Saddam Hussein, 9/11," 2009.

[37] The reference here to the Wolfowitz Doctrine derives from an early draft of U.S. Defense Planning Guidance for fiscal years 1994–1999, reputedly drafted by then–Undersecretary of Defense for Policy Paul D. Wolfowitz in February 1992, that was leaked to the *New York Times* and published by that paper. Although the final guidance took a somewhat different tone, the published draft excerpts are often credited as the first draft of an approach that Wolfowitz, as deputy secretary of defense, and others pursued toward Iraq and Iran from the beginning of the George H. W. Bush administration. See Tyler, 1992: "[T]he new draft sketches a world in which there is one dominant military power whose leaders 'must maintain the mechanisms for deterring potential competitors from even aspiring to a larger regional or global role.'"

[38] This high degree of predictability had been badly shaken at the time of 9/11, when it was not clear that the Al Qaeda attacks on the United States were essentially disconnected from developments in the Persian Gulf region and, more particularly, from Iran and Iraq.

[39] As discussed elsewhere in this work, the Europeans must also be concerned about the impact that turmoil in the Middle East and environs and the rise of terrorism could have on some members of the now-significant Muslim populations on the Continent.

[40] This point illustrates the fallacy of so-called energy independence as pursued by U.S. political leaders for more than a generation. Given that Western economies as a whole and, now and increasingly, the global economy depend on energy supplies from the Persian Gulf region, any one country's independence from oil exports from the region would not insulate that country from the consequences if other nations lost access to these supplies.

settled: whether the export of hydrocarbons from the region—which, for most countries in the world, is the only critical factor that matters regarding at least the Persian Gulf portion of the Middle East—is indeed at serious risk of physical interruption as opposed to commercial interruption (most notably, price management by the Organization of Petroleum Exporting Countries [OPEC]). There is little history of interruption, however, on which to base decisions. Even during the spate of nationalizations that began in the 1950s, the oil continued to flow. There was a declared Arab oil embargo during the 1973 Yom Kippur War (and, to a lesser degree, during the Six-Day War), but it is not clear that there was, indeed, an embargo that removed significant quantities of oil from the world market. That market is highly adaptable. To a great extent, oil is a fungible commodity, and, in 1973, there was a good deal of play-acting on this issue. Even after Iraq's destruction of the Kuwaiti oil fields in 1991, production was restored in relatively short order (despite predictions of a lengthy interruption). To be sure, the 2003 invasion of Iraq caused Iraq's oil production to tail off dramatically for a period, and its prewar production levels were not achieved until 2007, but the market (and alternative suppliers) successfully adapted.[41]

Nevertheless, potential turmoil in and around the region of the Persian Gulf is never reassuring. Although the 2003 invasion removed Iraq as a potential source of threat both to oil exports—a threat Iraq posed when it invaded Kuwait[42]—and in other senses (except to the degree to which Iraq became, after the invasion, a base for terrorists), uncertainties have remained about regional security. Some relate to the risk of terrorist action against pipelines (which are far more vulnerable than production facilities but also more rapidly repairable); some relate to possible Iranian activities that could jeopardize other nations' security; some relate to the continuing conflict between Turkey and the Partiya Karkerên Kurdistan [Kurdistan Workers' Party] (PKK) in Iraq and to the risk that Iraq's Kurds will declare independence; and some relate to the possibility of conflict between the United States and Iran and the potential for the latter to try shutting down oil-tanker transit through the Strait of Hormuz.[43]

Other concerns remain about Iraq's long-term future (including its possible return, one day, to the ranks of major regional powers); about the security of Arab states in the Persian Gulf (especially in regard to their relations with Iran but also in regard to their relations with one another); about the potential for the projection of terrorism from the region into other countries; about the continuing spread of Islamist radicalism (includ-

---

[41] Estimated Iraqi oil production was 2,116 thousand barrels a day in 2002; 1,344 thousand barrels a day in 2003 (which bridged the initial war period); and 2,145 thousand barrels a day in 2007. See Finlay, 2008.

[42] One of the continuing mysteries after Iraq's invastion of Kuwait in 1980 is why President Saddam did not continue the conflict in order to seize the oil fields of the Eastern Province of Saudi Arabia, which he most likely could have done before U.S. expeditionary forces were deployed to Saudi Arabia in strength.

[43] An Israeli attack on Iran, which is more likely than a U.S. attack, would be unlikely to focus on the Strait of Hormuz, but the disruptive effects of such an attack would still be mammoth, certainly in the region and, very likely, beyond.

ing from Saudi Arabia); about the impact of regional developments on the Arab-Israeli conflict (and vice versa); about the capacity of outsiders to do business safely in the region; about the impact of regional events on the large number of Middle Eastern Muslim expatriates, especially in Europe; and about the interests of other external players, including Russia, China, and India. In sum, the United States, its allies, and other countries cannot be indifferent to the state of security in the region, to the lack of a formal security structure, or to the severe limitations on today's system of security in terms of providing a reasonable degree both of predictability and of confidence about the future. This is the context that must now be addressed, and chief responsibility for doing something about it rests with the United States, if only because no other country is both able and willing to do so.

CHAPTER FOUR

The Core Challenges for a New Security Structure

For a region as complex at the Persian Gulf, with so many conflicting interests and so many different players—some of which are states, and some of which are nonstate actors inside or outside the region—it is difficult to present in a reasonably short form all of the requirements for a workable security structure. At the very least, there is the question, *Workable from whose perspective?* From the perspective of the United States, and of its friends and allies, the answer needs to be *our perspective.* Of course, that answer is clearly not good enough. For a security structure to work for the United States in a manner that goes beyond pure containment (or conflict-victory and post-conflict occupation), it also has to work for enough other active players so that few or none of them will have an interest, incentive, or ambition to knock it off its pins or that, even if they do attempt to undermine it, they will not be able to do so to the degree that the structure will collapse. In other words, for there to be a workable security structure, there needs to be a critical mass of members and supporters both inside and, where relevant, outside of the region that see the more-or-less effective functioning of the structure as being worth more than either continuing without it or working against it. They must also hold this view with sufficient strength both to keep potential spoilers from causing the structure to fail and to maintain a sense of the structure's worth that is sufficiently long term that it can come into being in the first place and pass at least its initial trials successfully.

Within that framework, from the United States' point of view, there are seven primary, region-specific parameters for a new security structure for the Persian Gulf region: the future of Iraq; Iran; asymmetric threats; regional reassurance; the Arab-Israeli conflict; regional tensions, crises, and conflicts; and the roles of other external actors. All are critical, but the first three discussed here—Iraq, Iran, and asymmetric threats—are likely to be more important than the others, at least for the next several years.

The Future of Iraq

It remains difficult, in early 2010, to predict with a high degree of confidence the possible course of the so-called endgame in Iraq—if that is, indeed, the proper name for the current phase of conflict—even though President Obama has set a timetable for withdrawing all but 35,000–50,000 U.S. troops by August 31, 2010, and even though the U.S.-Iraqi Security Agreement calls for the completion of the withdrawal by the end of 2011.[1]

The difficulty lies not just in the fact that the United States and all the other players in Iraq must make many decisions in the immediate future that could render outdated some tactical points made in this work. It also lies in the fact that it is far from clear just what the United States, as the lead nation, will want to see emerge from the current conflict in Iraq, much less what it will be able to achieve. It is also unclear what will happen in Iraq as U.S. forces are drawn down and whether some events could cause both the U.S. and the Iraqi governments to change course. For example, in July 2009, Iraqi Prime Minister Nuri Al-Maliki said that his government might ask the United States to retain troops in Iraq after 2011: "If Iraqi forces need more training and support, we will reexamine the agreement at that time, based on our own national needs."[2] Although this is not a definitive position, if such a decision were made, a significant shift in U.S. thinking, including about its posture and policies in the rest of the Persian Gulf region, would be required.

Indeed, despite the clarity of the United States' current course of action, there needs to be caution in regard to the analysis of several points. One such issue is what situation the United States should, in light of its own interests, want to emerge in Iraq versus what situation might reasonably be expected to transpire. Leaving aside what the United States might *like* to do in Iraq for broader strategic or other reasons, what its interests *should* lead it to desire are quite limited, beyond issues relating to the *manner* in which the United States and other Coalition forces change their situation in and with regard to Iraq. Most simply, at bottom, the United States has a two-fold interest in Iraq: first, that it not become a platform, either as a state or as a locus for nonstate actors, that poses threats to its neighbors (e.g., threats to Turkey caused by actions of Iraqi Kurds) and—in terms of Iraq's becoming a terrorist base—threats to Arab states of the Persian Gulf, Israel, and even Iran as well as to countries beyond the Middle East; second, that it not become so vulnerable to incursions from outside (notably, Iran, but also Turkey[3]) that this vulnerability intensifies the concern on the part of other regional states that their own vulnerability is increased.[4]

---

[1] Obama, 2009a.

[2] See United States Institute of Peace, 2009.

[3] Indeed, Iraq could also become vulnerable to an attack by Turkey. Unless, however, a Turkish incursion into Iraq—e.g., into the Kurdish areas—led to the general dismemberment of Iraq and to uncertainties about whether it would become a platform for exporting instability or importing instability (e.g., from Iran), it is not clear how much this would affect basic U.S. interests, unless conflict within Iraq escalated (e.g., to draw in Iran).

[4] The two criteria emphasized here for calculating basic U.S. interests in Iraq do not take account of the potential products of a civil war or of Iraq's division into more than one part.

Of course, this short list of potentially dangerous circumstances for the United States ignores the many political and moral issues that could emerge in full force if Iraq descends into civil war and that must be accounted for in any viable policy for Iraq's future. In addition, this list does not account for any plans the United States could develop—subject to Iraqi concurrence—to retain forces and bases in Iraq after 2011. There have already been some indications that such developments could be the course of U.S. policy.[5] These forces and bases could be designed to protect remaining U.S. nonmilitary assets (e.g., the embassy complex, aid agencies, and nongovernmental organizations [NGOs]), to engage in training Iraqi security personnel (which is critical to the possibility of achieving a reasonable level of stability and security in that country), and to inhibit incursions by outsiders whose activities could work against U.S. interests or those of others in the region that are important to the United States.[6] This last function requires retaining some capacity to counter an influx or resurgence of terrorists and to send a signal to outside powers—Iran, in particular—that a rise in influence that exceeded an imprecise but critical point would negatively affect U.S. interests in Iraq and in the region.[7]

Beyond that, there is the issue of whether the United States would like to retain bases and military personnel in Iraq for long-term purposes, including to deal with Iran, either in terms of deterring Iranian activities against Iraq (as discussed above) or elsewhere or in preparation for the possible use of military power against Iran. The last-named point relates to the choices that will be discussed in the next section of this chapter. Of course, the United States could seek to keep military bases and forces in Iraq simply as a geopolitical insurance policy or means of gaining influence (although some deployments could prove to be a double-edged sword) or to be able to deploy force rapidly to deal with other security situations in the region, including in the areas embracing and abutting Afghanistan.

Whether meeting the requirements for providing security for U.S. forces kept in Iraq for this purpose would be worth the risk, comparing risk to benefits would require careful analysis. Iraqi acquiescence in such a U.S. prophylactic strategy would be less

---

[5] For instance, see the following summary of a statement by General George Casey:

> The Pentagon is prepared to leave fighting forces in Iraq for as long as a decade despite an agreement between the United States and Iraq that would bring all American troops home by 2012, the top U.S. Army officer said Tuesday. Gen. George Casey, the Army chief of staff, said the world remains dangerous and unpredictable, and the Pentagon must plan for extended U.S. combat and stability operations in two wars. "Global trends are pushing in the wrong direction," Casey said. "They fundamentally will change how the Army works." (Curley, 2009)

[6] This list of possible U.S. interests suggests strongly that the security agreement will be modified before the end of 2011 or, at least, that the United States will seek such a modification unless there is a radical change in U.S.-Iranian relations for the better by that time.

[7] This last point begs the question whether those elements in Iraq with which Iran has a close association are those that are more likely to pose threats or other challenges either to departing U.S. troops or to the government in Baghdad rather than problem elements that could emanate from the Sunni community.

likely than its acceptance of a continued U.S. force presence for purposes directly benefiting Iraq. There is also the issue of whether U.S. forces remaining in Iraq would be part of any broader "guarantee force" provided by the United States (possibly with one or more European allies or other partners) to reassure various states in the region that it still retains interests and is prepared to be the arbiter of last resort in helping to forestall a recrudescence of conflicts, even after the creation of a new security structure. Iran is the regional country that would likely have the greatest problem with U.S. forces being present in Iraq beyond helping to stabilize it.

There is also the longer term to be considered in regard to Iraq. In terms of any form of regional security, and, certainly, any form of security structure, Iraq is, at the moment, a nonplayer and is likely to remain so for the considerable future, despite its plans to acquire modern equipment, such as tanks and advanced fighter jets.[8] But it will not always be thus. At some point, full practical sovereignty will revert to Iraq, in whatever form or configuration that country has taken, and, in time, Iraq can be expected to build serious military capacities. At that point, Iraq can also be expected to become a part of regional politics, economics, and security; it is also likely, under one form or another of domestic political leadership and one form or another of territorial and political organization, to become again a contender for power and influence in the Persian Gulf region. It would be a historical anomaly if this did not happen. Furthermore, Iraq is unlikely to be proof against meddling from one or more of its Arab neighbors—notably, Syria and Saudi Arabia—and Iran. Thus, its government will likely have an interest in being included in any new security structure, both for its own sense of pride and statehood and as a hedge against encroachment, even from a Sunni neighbor. Thus, Iraq should be included in any political process being developed regarding regional security.

It should be clear that the process of dealing with Iraq in the period immediately ahead would be facilitated, for all and sundry, if the foundations were being laid for a new security structure in the Persian Gulf. Indeed, it is possible that at least some of the parameters for that structure should be set precisely as tools to deal with Iraq's near-term future. These parameters could then be expanded in order to develop a more-encompassing system that would also deal with the full range of other considerations.

There are two broad cases for Iraq's future in this regard. In one, the government in Baghdad could become interested in such a process and could deliver at least a significant fraction of the political players and interests in the country, if only on the basis of seeking to create conditions in which Iraq would have a greater, rather than lesser, capacity to decide its own future, without negative interference from another party. Of course, perceptions of this interest on the part of Baghdad will differ from group to group, sect to sect, and region to region, but the degree of coming together might be sufficient for the country to decide to participate in a larger effort that would try

[8] See, for instance, Reid, 2009.

to secure for Iraq a significant measure of capacity to shape its own future within the region, however its long-term internal political processes develop.

In addition, the most-important external regional influence on the immediate future of Iraq—namely, Iran—would have to see that its interests are more likely to be secured through a process that seeks to create a framework for an orderly transition of foreign forces out of Iraq and some means of isolating (if not resolving) its continuing internal struggles than through either a laissez-faire approach or an effort to gain tactical advantage in Iraq, even at the price of promoting, rather than retarding, turmoil. Whether Iran would see advantages in a relatively hands-off approach to Iraq, provided that there were some clear understandings with other parties, including the United States, can only be determined through a diplomatic process. That process might be facilitated if it were conducted against the background of practical ideas for a security structure within which Iraq's near-term future would be considered.

The alternative case is that there could be insufficient prospect that the Iraqi government would be able or willing to seek support from an encompassing security structure, that Iran would be willing to take part, that the United States would be willing to accept the required limitations on its own freedom of action, or that all other relevant countries and nonstate entities were prepared to abstain from playing a spoiler's role. In fact, that all of these pieces will come together as desired may be too much to hope for. Even so, it is still worthwhile to make the effort to determine whether the Iraqi endgame can be imbedded in a process and structure in a way that affords at least some hope of success to all who place a higher value on reduced violence and uncertainty than on their ability to retain the full capacity for free play in the future.

To meet this goal, efforts to work toward at least a rudimentary security structure would need to consist of at least the following elements, some of which should be followed in any event:

- bilateral and multilateral diplomacy involving all the parties relevant to Iraq and including a commitment from all parties to take part in creating and subscribing to a set of basic goals for Iraq, as acceptable to the Iraqi government, and to accept important self-denying ordinances (e.g., on the use of force by outsiders to affect events in Iraq).
- a regional conference as either a cover for or ratification of diplomacy. Authority for other efforts would derive from the conference, which, ideally, would continue to exist and involve officials meeting in permanent session.
- participation in multilateral diplomacy by the United States, Iraq, Iran, Turkey, the GCC states, and the UN. Others (such as representatives of the EU, individual European and Middle Eastern countries, the Russian Federation, Japan, China, and India) should be invited to take part as appropriate and by common consent.

- Iran's agreement that it would prefer to see an orderly reduction and departure of U.S. forces from Iraq and the development of effective Iraqi security forces rather than a chaotic situation; appropriate behavior would also be required on the part of Iranian-influenced elements in Iraq.
- continuing reiteration by the United States of its future preferences regarding Iraq, including the possibility of retained U.S. bases and forces, with transparency about these preferences in discussions with other parties. Even if the Iraqi government asks the United States to keep forces in the country after 2011, for one purpose or another, the United States needs to make clear its own long-term aspirations.[9]
- a joint commitment from, and practical efforts by, all regional parties to oppose terrorism in all forms and from all sources throughout the applicable territory of the Persian Gulf region and beyond (e.g., in Afghanistan and Pakistan). Of course, if the "applicable territory" were extended to include Israel, Palestine, and Lebanon, an extension of this joint commitment would also be required.
- the creation of a Standing Military Commission, composed of all the core members of multilateral diplomacy listed above, run under Iraqi leadership. Its purposes would include agreeing on definitions regarding the military situation in Iraq as it evolves; developing possible limitations on outside involvement (to be decided by the plenary conference); creating a system of inspections; and devising confidence-building measures (CBMs), including exchanges of information on military issues.
- the creation of a Standing Political Commission, with the same membership as the Standing Military Commission and also run under Iraqi leadership. Its purposes would include discussing and clarifying the interests of all the different parties, exchanging statements of requirements for building political confidence in each another's activities and intentions, developing tools for dealing with terrorism, and creating teams of experts to assess Iraqi material requirements (e.g., reconstruction, development). All would be a prelude to any future donors' conferences designed to organize both bilateral and multilateral support for Iraq.[10]
- development of the concept of the Standing Military Commission and the Standing Political Commission for use in a regionwide security structure.

---

[9] The United States has already said that it does not want to retain forces or bases in Iraq. This provision is included in the event that circumstances change. See President Obama's comments at Camp Lejeune:

> And under the Status of Forces Agreement with the Iraqi government, I intend to remove all U.S. troops from Iraq by the end of 2011. We will complete this transition to Iraqi responsibility. . . . So to the Iraqi people, let me be clear about America's intentions. The United States pursues no claim on your territory or your resources. We respect your sovereignty and the tremendous sacrifices you have made for your country. (Obama, 2009a)

[10] A donors' conference for Iraq was held at Madrid on October 23–24, 2003. See Pan, 2003.

In any event, as the U.S. draws down its force presence in Iraq, economic and political reconstruction and development will be critical efforts. Whether or not Iraq would play an important role in any regional security structure, it is a real country with real people whose situation cannot simply be written off as outside the perspective of calculations made on the basis of realpolitik. Although Iraq is no longer a poster child for what might be done in terms of democratization, supporting indigenous efforts to promote better governance vis-à-vis human rights and helping the Iraqi people to live better lives is not just a moral and a political issue: In the fullness of time, such support will be a key element of helping to shape societies in order to help promote security in the broadest sense of the term. Thus, the endgame in Iraq will, of necessity, be expensive in terms of the role of outsiders in helping it rebuild its society, however it is organized politically, and even beyond Iraq's ability to use its own oil revenues for its reconstruction and development: Such resources are most unlikely to be sufficient for the task, at least during the next several years.[11] Outsiders—e.g., other governments, such international institutions as the World Bank, the private sector, NGOs—will be needed to do much of the work, whoever foots the bill. And much of this effort can and should come from the United States' European allies as well as from Iraq's wealthy, oil-producing neighbors. Some of that effort should be military, as is the NATO Training Mission–Iraq (NTM-I); some should be denominated in terms of Iraq's long-term relations with the EU; and some should entail both large-scale, European-led economic- and government-support efforts designed to buttress local activities and the engagement of other outside states and international economic and financial institutions, especially the UN and its affiliated agencies.

Clearly, how the United States moves beyond its current engagement in Iraq and how and to what extent other countries become engaged will also affect the way in which the United States is regarded by other powers in the region, both those that look to the United States for support, especially for security, and those that see the United States as a competitor or that harbor hostile attitudes and intentions toward any U.S. or other Western presence in the region. Most important in this latter category, at least at the moment, is, of course, Iran.

Iran

For the past 30 years, the United States has been preoccupied with the situation, status, role, ambitions, policies, and practices of the Islamic Republic of Iran. This will continue to be true for the foreseeable future whether the Iranian regime persists in its current policies or is replaced by a regime that is more amenable to the United States

[11] Iraq currently has the world's third-largest amount of proven oil reserves, estimated at the end of 2008 at 115 billion barrels. Its capacity to pay for its reconstruction and development is, thus, a matter of production rather than of finding oil to sell abroad. See BP, 2009, p. 6.

and its interests. This latter qualification is important. There is often a too-easily-made assumption that the nature of a nation's government has a decisive impact on how it comports itself in the outside world. Sometimes that can be true, as in the cases of Nazi Germany and the Soviet Union, although, in both of those states, national ambitions toward the outside world derived, at least in part, from history, geography, and economics, not to put too much weight on the question of national culture. Even the argument that democracies do not make war on other democracies is not conclusive, although there are many examples that seem to validate the proposition.[12]

In Iran's case, it is not at all clear that the passing of the Islamic government or regime would lead the country to play a role in the region that the United States and others would find sufficiently noncompetitive and, of course, sufficiently nonthreatening. This point is particularly important to consider at the moment because of the turmoil that followed the June 2009 elections in Iran and because of hopes expressed in the West that a major transformation in the Islamic Republic, its governance, and its policies is in the offing. Indeed, although Iran has a better record than many other countries, including many in the Middle East, in terms of not having pursued territorial conquest against neighbors[13] during at least the last two centuries, it is still an ambitious country with a strong sense of pride and of what it sees to be its rightful place in the front ranks of nations in its neighborhood and beyond.[14] (See Figure 4.1 for a map of Iran and its environs.) That does not mean that, after the end of the era of the Islamic Republic, Iran would, necessarily be aggressive or, short of that, structurally uncooperative with the West. Even if democratic, however, it would be unlikely to be quiescent in terms of abstaining from seeking to play an active or even major role in the region. It could wish to be a hegemonic nation in the region, although one must always be careful in invoking the term *hegemonic*: Many countries that want to be hegemons in their immediate vicinity are unable to indulge that ambition (or fan-

---

[12] Both NATO and the EU retain value, in part, because they serve as not-formally-acknowledged-but-still-widely-understood-as-such insurance policies regarding the future of Germany, even though Germany is now a solid democracy. Indeed, the EU, since its earliest institutional manifestation as the European Coal and Steel Community in 1950, was created not principally for economic reasons but rather to make war between Germany and its neighbors more difficult than in the past, and, hopefully, impossible. It was not for nothing that German Federal Chancellor Helmut Kohl, in supporting enlargement of both NATO and EU membership, sought that the first members admitted be Poland and the Czech Republic, whose membership would then "surround" Germany with NATO and EU members. Giving up the deutschmark in favor of the Euro also reflected, at least in part, a similar German geopolitical motivation.

[13] The UAE would dispute this characterization because of Iran's occupation of the three Persian Gulf islands of Abu Musa and the Great and Lesser Tunbs. The merits of the argument, historically, rest more with Iran than with the Arab states, but, nevertheless, this remains a bone of contention.

[14] Indeed, much analysis of the Iranian nuclear-development program fails to note that the program began under Shah Mohammad Reza Pahlavi and has strong popular backing because it allows Iran to demonstrate that it is a modern, technologically competent country. Thus, outsiders' proposals to prevent the construction of nuclear weapons would face much-stiffer domestic political opposition in Iran if they also tried to stifle Iran's development of civilian nuclear power.

Figure 4.1
Map of Iran

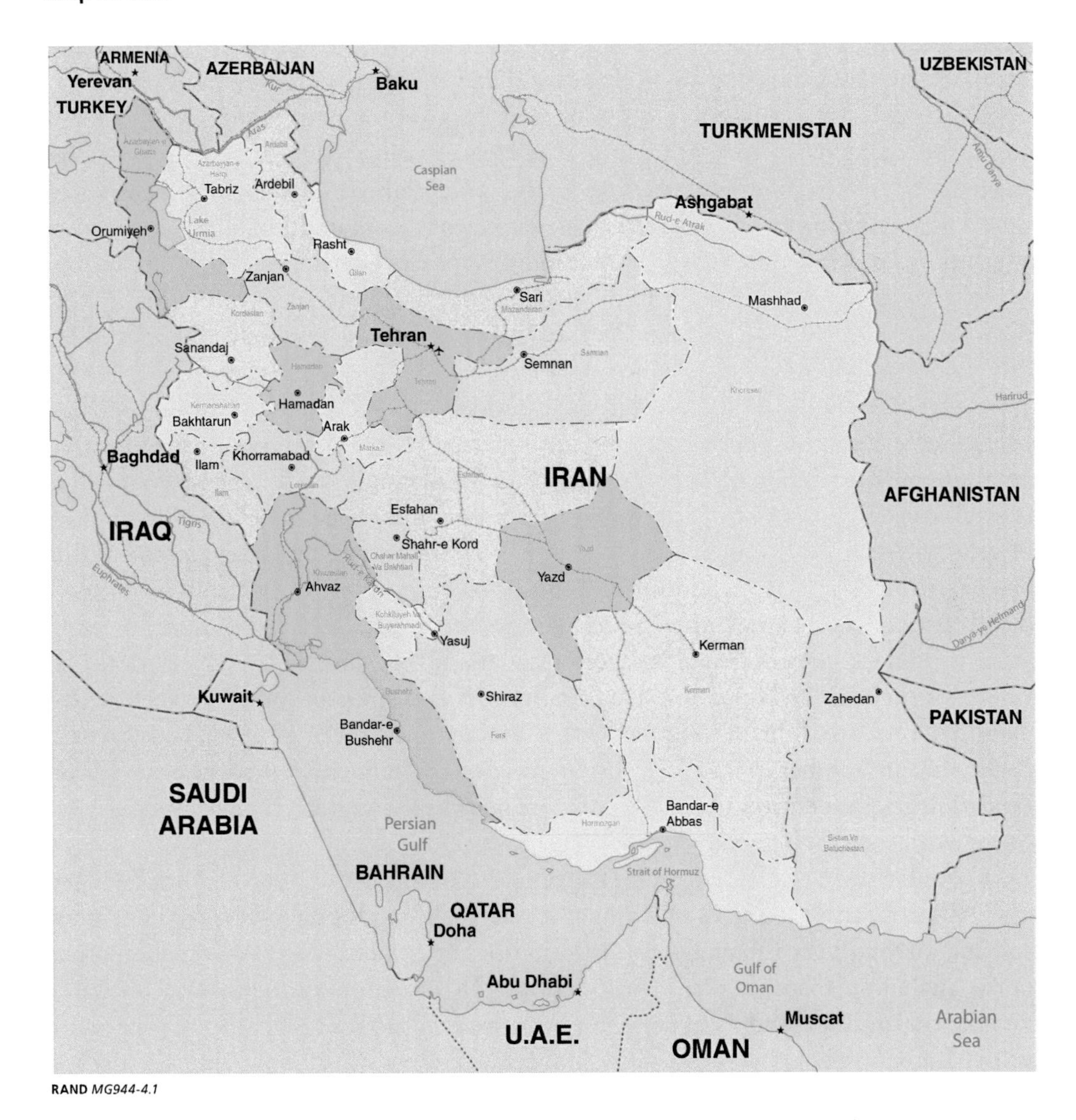

RAND *MG944-4.1*

tasy) either because the price they would have to pay would prove too great or because another power or powers are pushing back. In this case, leaving the Persian Gulf open to the fulfillment of any Iranian hegemonic ambitions would require not just that all other countries in the region be supine but that the United States abstain from engagement virtually altogether and that no other power, from the West (i.e., Europe) or the East (i.e., China, India, or Russia) would seek to play a significant role in the Persian Gulf. It is so unlikely that all these conditions would be fulfilled that it is safe to dismiss this scenario.

For the purposes of analysis, it is both useful and important to consider two cases in regard to Iran's future. They are not mutually exclusive, and they do not completely bound the range of possibilities; further, which of the two alternatives comes to dominate in shaping Iranian behavior, both internal and external, will depend, in part (but, as argued earlier, only in part), on the nature of its government or regime at any point, on the relationship of its government to society (e.g., in terms of the role of the religious establishment in temporal governance), on the potential role of the Iranian military, and on the actions of outside powers or nonstate entities that either do or can have a significant impact on Iran and its orientation toward the outside world.

These cases can be summarized as follows from the perspective of the United States and other Western powers: (1) Iran continues to be relatively uncooperative with major parts of the outside world or (2) Iran becomes relatively cooperative. In the former case, the reasonable approach for the United States to take, depending on what *uncooperative* means in practice, can be summarized as continuation of some form of containment. In the latter case, Iran could be a valid candidate for membership in a regional security structure. Failing that, there might at least be some accommodation between Iran and the United States (and other Western nations), a situation that, at the time of writing, the Obama administration was seeking to pursue. Of course, in cases when it is not easy for the United States (or another nation) to classify countries with which it interacts into the rigid categories of "friend" or "foe," a mixed relationship may be all that can be hoped for. This duality can make pursuit of a productive relationship in one area difficult to square with a competitive or even somewhat conflictual relationship in another area, especially because of the tendency of parts of the U.S. population and leadership, including some members of Congress, to have difficulty in dealing with such ambiguity.[15]

For some time, the United States has operated on the assumption that Iran will be largely uncooperative in terms of regional security, although potential exceptions may arise and although the Obama administration has attempted to explore an alternative course, including through direct negotiations. This assumption is derived, in part, from the continuing standoff in regard to Iran's nuclear program; it is also a compound both of Iranian statements—especially some by President Mahmoud Ahmadinejad—and of Iran's actions.[16] The latter category includes its support for Hezbollah in Lebanon and, to a lesser degree, its support for Hamas in the Palestinian territories (especially Gaza). Assessments that lead to a belief that Iran will tend to be uncooperative have been reinforced by the manner in which the ruling elite in Iran dealt with protests over the conduct of the 2009 presidential election.

---

[15] One only has to cite the often-difficult relations between the United States and France to see this tendency. Recall, for example, the way in which french fries were renamed "freedom fries" by the U.S. House of Representatives restaurant during the squabble over French and U.S. differences regarding the invasion of Iraq in 2003.

[16] These include President Ahmadinejad's comments about the Holocaust and the legitimacy of Israel as a state.

Of course, the record, even as viewed by skeptical U.S. officials, is not all one-sided: Iran played a role in supporting the U.S.-led military operation against Iraq during Operation Desert Storm in 1991, and it also supported the United States (and the Coalition) during its efforts to oust the Taliban from Afghanistan in 2001 and during the diplomatic efforts that followed, including the Bonn Conference[17] agreement of December 2001 and afterward. Certainly, Iran acted out of self-interest, but there is nothing exceptionable about that in statecraft. Indeed, cooperative relationships of at least a limited nature can often be built on the shared self-interest of parties that otherwise are at serious odds. Following this Iranian cooperation in Afghanistan, it was the United States that broke off contacts and foreclosed the exploration of possibilities, notably when President George W. Bush declared Iran to be a member of a so-called axis of evil.[18]

At the same time, as the United States conducts its drawdown of forces in Iraq, it is still not entirely clear whether Iran will play a role that could be characterized as either cooperative or uncooperative. On the one hand, it is obviously pleased to see the United States reduce its presence in Iraq, not least because of the proximity of U.S. military forces to Iran. At the same time, Iran needs to be wary of the development of an Iraq that is in turmoil or that in some other way is unsettling, even threatening, to Tehran. Being supportive of U.S. efforts to promote long-term stability in Iraq could, thus, on balance, be in Iran's interest. On the other hand, Iran might calculate that the period of the U.S. drawdown is an ideal time to try increasing its influence in Iraq, to the extent it is able to do so (i.e., in relationship to Iraq's Shi'a population and leadership), in part to pursue its overall regional ambitions. It could, thus, choose to try ramping up its capacity to meddle in Iraqi politics, to its own advantage and that of Iraqi factions friendly to Iran and, possibly, even at the expense of the current Shi'a-led government in Baghdad. Indeed, if the government in Baghdad proves unable to stay in power and turmoil does ensue, it cannot be ruled out that Iran would intervene militarily, directly or indirectly, to try securing Iraq against, for example, a return of Sunni political dominance.[19]

---

[17] See Agreement on Provisional Arrangements in Afghanistan Pending the Re-Establishment of Permanent Government Institutions, 2001. Also, see the following report of a presentation by the former U.S. special envoy to Afghanistan, Ambassador James Dobbins:

> Dobbins said Iran played a constructive role in the Bonn conference by suggesting that the agreement contain phrasing calling for democracy and fighting terrorism by the future Afghan government. Dobbins said his instructions at the Bonn conference were to press for "a broadly based, representative government. That was our objective. The word democracy was actually introduced into the Bonn talks on the recommendation of the Iranian delegation." (Kurata, 2005)

[18] President George W. Bush said, "States like these [Iraq, Iran, and North Korea], and their terrorist allies, constitute an axis of evil, arming to threaten the peace of the world" (G. W. Bush, 2002a).

[19] It also cannot be ruled out that Iran would intervene militarily in Iraq if Turkey sent major military forces into Iraqi Kurdistan, thereby threatening overall stability in the country and, perhaps, more broadly.

The answer to the basic question whether Iran will cooperate or not will likely emerge during 2010, and it is also likely to be affected by what else is happening—or not happening—in U.S.-Iranian relations. This is not to assume that Iran has the capacity to make life very difficult for the United States as the latter reduces its forces in Iraq. After all, a large part of the reemerging violence in Iraq is not perpetrated by Iraqi Shi'a, although this begs the question of the extent to which Iran could stir up trouble that would threaten either U.S. forces or U.S. political goals in Iraq. Yet, from the United States' point of view, an Iran that has a basic interest in a stable Iraq and the peaceful exit of U.S. forces is certainly preferable to one looking to make life uncomfortable for the United States in response, for example, to a new round of economic sanctions imposed on Iran because of the nuclear issue.

At the same time, there is less uncertainty about the possibility of Iranian cooperation, tacit or explicit, with the United States and other members of the International Security Assistance Force (ISAF) in Afghanistan. Although Iran's interests in Afghanistan—including its desire for influence there, especially in areas historically engaged with Iran, directly or indirectly—are unlikely to be congruent with those of the West, there is some clear correspondence of interests, including opposition to both the Taliban's return to power and the capacity of Al Qaeda to operate from Afghan (or, indeed, Pakistani) sanctuaries. In this sense, Iran will benefit, at least in the near term, from ISAF (including U.S.) military deployments and from outsiders' nonmilitary involvement. How that fact might translate into positive cooperation is another matter, however, especially given that the areas of Afghanistan where Iran has the greatest degree of influence are areas that are less under threat from the Taliban and Al Qaeda than are areas in the south and east. In this situation, too, however, there is potential either for a positive U.S.-Iranian relationship that could be built upon within the context of a diplomatic process or for Iran's making life more difficult for the United States (although its capacity to do so in Afghanistan is lower than in Iraq, given the axes of Iranian influence) if the overall U.S.-Iranian relationship took a more-conflictual course.

Suffice it to say, it is important that possibilities for U.S. and Iranian accommodation not be ruled out. Of course, the United States will want to "verify" in tandem with extending any "trust."

Assessments concerning the difficulties in the United States' seeking accommodation with Iran are less ambiguous in most, but by no means all, of the Arab states of the Persian Gulf. Indeed, for most of them, a key criterion for judging the U.S. drawdown from Iraq and other U.S. policies in the region is whether the United States remains willing and able to thwart any Iranian ambitions, not just toward Iraq but also in the region of the Persian Gulf in general. Such Iranian ambitions are expressed not just (or even primarily) in military terms but rather in terms of political, economic, and cultural (in some cases, religious) influence. By contrast, there are also concerns that the United States could seek an accommodation with Iran, potentially at the expense—

indeterminate, but still feared—of regional Arab states. Depending on how it goes, the diplomatic process with Iran begun in October 2009 by the United States, Britain, France, Germany, Russia, and China, along with a EU representative,[20] could fan the embers of any such concerns. As in all such ventures that are designed, in major part, to build confidence, transparency is a critical element of all related diplomacy. This is all the more important in a region where suspicion is endemic and conspiracy theories are a parlor game.

Of course, attitudes are not all one-dimensional. Oman, for instance, has had continuing relations with Iran and has, in general, sought to keep from being engaged in the anti-Iranian politics of most other Persian Gulf Arab states.[21] Some other regional states—even Saudi Arabia, which has been careful to ensure that it has a capacity to deal with Iran when particular events and situations merit such a course of action—have worked out various modi vivendi with Iran. Bahrain has worked to improve relations with Iran despite historic tensions related to Iran's earlier claims to the island nation (which Iran voluntarily renounced in 1970). In particular, in 2008, Bahrain went on record as seeking "a peaceful solution to the Iranian nuclear file, to avoid the scourge of war, and to enhance world peace and stability."[22]

The apposite point in the context of this work, however, is whether Iran would be both willing and able to play a constructive part in a regional security structure—again, *constructive* as defined by the United States, its allies, and its partners in the Persian Gulf.[23] Of course, Iran's neighbors would have to be prepared to reciprocate, which is not a foregone conclusion. Further, this proposition begs the question of what *kind* of security structure is being envisioned. Iran has, for some time, been willing to consider security arrangements with regional countries and has even proposed a broad-ranging structure—but usually with the proviso that it be composed solely of regional states without participation from outsiders, by which it means the West and, notably,

[20] Solana's mandate came to an end on December 1, 2009, as the Lisbon Treaty came into force. He has been succeeded by Baroness Catherine Ashton, who assumes the newly created office of EU High Representative of the Union for Foreign Affairs and Security Policy.

[21] For example, Sultan Qaboos bin Said Al Said of Oman paid a three-day official visit to Iran at the beginning of August 2009—the first visit of an Omani king to Iran in 35 years. There, "Oman and Iran signed a security agreement . . . [and held] talks for a gas swap deal which would see Oman supplying Iran with gas" (The Royal Forums, 2009). Notably, the visit had been announced in the midst of the political turmoil in Iran. See "Sultan Qaboos to Visit Iran," 2009.

[22] See Al-Khalifa, 2008. The Bahraini minister also referred to outstanding issues with Iran, including the status of the three Persian Gulf islands of Abu Musa and the Greater and Lesser Tunbs.

[23] The qualifier *able* is included here because it is possible that, in the period just ahead, the Iranian government may not be functionally capable of making decisions of the complexity and magnitude of those involved in the creation of, and Iran's accession to, a regional security structure. This potential inhibiting factor relates most obviously to what may transpire internally in Iran following the 2009 Iranian presidential election and its aftermath. Paralysis of government, even regarding national-security issues in Iran's palpable self-interest, could ensue.

the United States.[24] The qualification "usually" is important. In its proposals for dialogue in September 2009, Iran used a nuanced formulation:

> The Iranian nation is prepared to enter into dialogue and negotiation in order to lay the ground for lasting peace and *regionally inspired and generated* stability for the region and beyond and for the continued progress and prosperity of the nations of the region and the world.[25]

Iran also regularly calls for a Middle East that is free of weapons of mass destruction (WMDs). This clearly refers to Israel, but it could also refer to Pakistan,[26] which,

---

[24] See, for instance, the following ten-point proposal made to the World Economic Forum in Doha, Qatar, in April 2007 by Hassan Rowhani, who had been Iran's chief nuclear negotiator under President Mohammad Khatami:

> 1. Establishment of a Persian Gulf Security and Cooperation Organization comprising the six member states of the Gulf Cooperation Council (GCC) as well as Iran and Iraq in accordance with Clause 8 of Resolution 598 of the United Nations Security Council.
> 2. Preparing common security grounds for fighting terrorism, organized crime and drug smuggling, as well as other joint security concerns.
> 3. Gradual removal of all restrictions in political, security, economic and cultural fields.
> 4. Development of trade ties by taking the countries' potentials into consideration and conducting joint investment in economic projects to achieve a regional free-trade mechanism.
> 5. Guaranteeing the security and energy export of regional countries to secure their interests and achieving a sustainable mechanism for energy needed by the world.
> 6. Building confidence among regional countries in the nuclear field.
> 7. Setting up a joint consortium for uranium enrichment among regional countries to procure nuclear fuel and other peaceful nuclear activities under the supervision of the International Atomic Energy Agency.
> 8. Forging serious cooperation among regional countries for having a Middle East free of weapons of mass destruction.
> 9. Putting an end to arms races in the region by providing resources for the purpose of economic development and fighting poverty.
> 10. *Making foreign military personnel exit the region and establishing full security by the regional countries.* (Rowhani, 2007 [emphasis added])

See also "Proposals for Persian Gulf Security: Build Trust, Cooperation," 2007. However, as Afrasiabi, 2007, notes,

> Unfortunately, there is only a dim prospect for this proposal's acceptance by the Arab states of the GCC, which have devised their own version of "collective security" that does not include the region's two most populous states, Iran and Iraq, and which have traditionally relied on U.S. protectorate power and are therefore averse to any security plan that might actually increase their sense of vulnerability *vis-à-vis* their assertive non-Arab neighbor, Iran.

[25] Islamic Republic of Iran, 2009 (emphasis added).

[26] Iran may also have concerns about Indian ambitions in the Persian Gulf region, concerns that may have been augmented by the July 2009 agreement between the United States and India that presages significant U.S. arms sales to India. See "India, U.S. Agree on End User Monitoring Pact," 2009.

although not technically a Middle Eastern country, is one about which Iran has serious security concerns. Thus, in its September 2009 proposals for negotiations, Iran endorsed

> [p]romoting the universality of NPT [Nuclear Non-Proliferation Treaty] global resolve and putting into action real and fundamental programmes toward complete disarmament and preventing development and proliferation of nuclear, chemical and microbial weapons.[27]

Notably, this formulation would also apply to Iran, a point very much apposite to debates about its willingness to negotiate regarding its own nuclear program.

Of course, Iran may not be prepared to see its long-range interests as being best served by regional security arrangements that would place the qualities of stability and predictability, however defined, above its own freedom of action to pursue its regional ambitions. These arrangements would be based on a structure that would not, perforce, exclude outsiders. If Iran is not prepared to accept these arrangements, then a further question would have to be asked: Is it possible to craft any structure that, with Iran's abstention and likely active opposition, would be able to meet a sufficient range of criteria to succeed, including the ability of the United States both to reassure regional countries of its commitment to security and to take a step back and work, over time, toward a largely over-the-horizon military presence, with capacity to reinsert forces when and if needed? *This question about Iran's attitudes and role may be the most-problematic question of all in regard to the potential for creating a viable Persian Gulf regional security structure.*

In the design of a new security structure, therefore, both cases—Iran opting in and Iran opting out—would need to be provided for. Indeed, given that the development of a new structure would almost certainty take place in stages over a considerable period, that structure's character at every point would need to account for a potential evolution of Iran's calculations about its own interests and for the possibility that Iran would, over time, decide that it could best balance those interests by joining efforts being undertaken to provide greater assurance regarding reductions in the risk of conflict. (Certainly, other members' willingness to accept Iran's participation in the security structure would be indispensable.) Indeed, it could very likely be that the evolution of a new security structure, absent Iran, that was proving to be successful in terms of the potential for regionwide cooperation could provide incentives for Tehran to modify its policy of abstention if it found that it could not confound such arrangements. Of course, this possibility would depend on the development of a security structure that was not premised, from the outset, on Iran's being the enemy or, at least, the outsider against whom the security efforts of members of the new structure would be directed. This is clearly not an easy distinction to draw. It may be that, as with the develop-

---

[27] Islamic Republic of Iran, 2009.

ment of both the Western and the Eastern security structures in Europe during the Cold War, an essential requirement will prove to be the existence of an "other" against whom to rally support among the rest.

Therefore, a critical question, which cannot be answered except in light of experience, is whether something on the order of the post–Cold War efforts to craft a new security structure in Europe could be attempted in regard to the Persian Gulf. In other words, would it be possible to craft a structure that is not premised on the existence of an enemy (or enemies, as was this case in the creation of both NATO and the Warsaw Pact) but rather on a situation of potential instability or uncertainty that could function as the common enemy?[28] Could there be sufficient common interest on the part of all key players (in this context, the regional countries) in pursuing such a set of relationships rather than risking a reversion to the tensions and potential conflict of the earlier period? Of course, a security structure that has to be premised on the existence of a hostile "other" is, perforce, of much more-limited value for the long term than is one that offers, from the beginning, the possibility of universality. Indeed, a security structure that would require in Iran the kind of evolution that took place over more than 70 years in the Soviet Union (and eventually led to the end of the Cold War) might be all that is possible, given the nature of Iranian political society (and of other countries in the region). But, from the perspective of the goals that inform this work, that would be a decidedly second-best choice.

Obviously, the kind of security structure contemplated for the region of the Persian Gulf will differ radically depending on whether Iran were viewed, rightly or wrongly, as the odd man out (at best) or the enemy (at worst) and on whether a place for Iranian participation in the structure, early or later, were provided (if only through not naming Iran as the target of security arrangements).[29] This does not mean that Iran

---

[28] Some commentators have questioned whether NATO can remain relevant after the end of the Cold War without an enemy. The simple answer is that it remains, at the very least, an insurance policy against a common concern (i.e., an enemy): uncertainty about the future. Given Europe's bloody history in the first half of the 20th century, which was followed by the extraordinary risks of the Cold War period, this is not an unworthy or trivial objective for the Atlantic Alliance, even if it had no other purposes or tasks to perform.

[29] The preamble to the Brussels Treaty of March 17, 1948, specifically mentioned Germany as a potential enemy: "[t]o take such steps as may be held to be necessary in the event of a renewal by Germany of a policy of aggression" (Treaty of Economic, Social, and Cultural Collaboration and Collective Self-Defense, 1948). The modified Brussels Treaty, through which West Germany and Italy were admitted into what was then to be called the Western European Union, replaced that clause with the following: "to promote the unity and to encourage the progressive integration of Europe" (Modified Brussels Treaty, 1954). By contrast, the North Atlantic Treaty was clearly understood to be directed against the Soviet Union but was silent on the existence of any enemy (North Atlantic Treaty, 1949). For its part, the Warsaw Pact did not mention an enemy, although it did make reference to actions by the West that, purportedly, were the motivation for the agreement:

> [T]he formation of a new military alignment in the shape of "Western European Union," with the participation of a remilitarized Western Germany and the integration of the latter in the North-Atlantic bloc . . . increased

should simply be given a free pass to become part of the wider security structure at any point and under any circumstances it chooses. Indeed, for any such structure to have a chance of success—other than, say, by being based on the continuing, active presence of strong U.S. military power and a simple strategy of containing Iran—there would need to be criteria for the structure itself, beginning with the ideas presented earlier in the case of Iraq, and for any country to be accorded membership, formal or informal.

In the case of Iran, the United States (for certain), several of the regional Arab states, and some European states would require that some specialized criteria, including the following, be met:

- adequate resolution of issues regarding Iran's nuclear-development programs, especially the establishment of concrete steps for determining (i.e., verifying) that Iran is neither seeking to develop nuclear weapons nor even, for the purposes of building confidence among the states, nearing what is commonly called a *breakout capability*. However, it may already be too late—or nearly too late—to prevent Iran from reaching the latter point if it chooses this course.
- Iranian abstention from efforts to make more difficult the resolution of key issues related to both security and politics in Iraq's immediate future
- Iranian willingness to support efforts to stabilize Afghanistan or, at least, an agreement not to make matters worse for the United States and NATO
- Iranian willingness to seek positive relations with its Persian Gulf neighbors if they are willing to reciprocate on the basis of commonsense standards
- Iranian willingness to abandon its support of any individuals or organizations that practice or support terrorism
- Iranian willingness to support the Arab-Israeli peace process or, at least, not actively to interfere in or oppose diplomatic efforts to resolve the conflict in any of its key particulars, including Israeli-Palestinian relations.[30]

Many of Iran's neighbors and other countries in the region have, over the years, desired an end to Iran's efforts to export its revolution to other states. How serious these Iranian efforts are at this time is open to question, however, although there are mixed feelings in some Gulf Arab states (e.g., Dubai) about Iran's commercial influence and the large number of Iranian expatriates, factors that provide trade advantages on the

---

the danger of another war and constitutes a threat to the national security of the peaceable states. (Treaty of Friendship, Cooperation and Mutual Assistance Between the People's Republic of Albania, the People's Republic of Bulgaria, the Hungarian People's Republic, the German Democratic Republic, the Polish People's Republic, the Rumanian People's Republic, the Union of Soviet Socialist Republics and the Czechoslovak Republic, 1955)

[30] In September 2008, President Ahmadinejad said the following: "'If they [the Palestinians] want to keep the Zionists, they can stay . . . . Whatever the people decide, we will respect it. I mean, it's very much in correspondence with our proposal to allow Palestinian people to decide through free referendums'" (quoted in Tatchell, 2008).

one hand but a risk of rising Iranian influence on the other. There is also a question about the importance of such activity and such influence in terms of fundamental requirements for the effective management of a regional security structure. Indeed, this point touches on a central issue: Creating a security structure does not mean an end to all differences, all strife, all competition, and all the activities of one country that could be objected to by another. If all such conflict were eliminated, it is very likely that there would be no need for a security structure in the first place. The objective of such a structure and the basic understandings that would buttress it is to get every party to agree that it will not seek to reap net gains from certain negative actions, especially actions that either entail armed conflict or could lead to armed conflict or other forms of hostile action, including subversion of legitimate governments, attempts to foment insurrection, and support for insurgencies or terrorism. If successful, *the security structure would serve as a sort of firebreak* to help prevent normal stresses and strains from reaching the point at which it becomes difficult or impossible to prevent escalation to a level of tension or conflict at which none of the parties gains and all potentially lose. This is a cost-benefit analysis that concludes that, in the common interest, certain behavior should be ruled out of bounds to keep it from spiraling out of control.

At the same time, of course, Iran would have its own criteria either for taking part in the regional security structure or, at least, for abstaining from trying to confound a structure from which it had either been excluded or had chosen to exclude itself. These criteria could include

- security guarantees to Iran, underwritten by the United States and others and infused with a high degree of credibility, that were contingent on Iran's meeting security and other reasonable requirements (especially regarding the Iranian nuclear program, but also regarding terrorism, Israel, and the subversion of regional states or governments) propounded by the United States and others
- an end to economic sanctions, both unilateral and multilateral, and a full reintegration of Iran into the global economy and political commerce
- an end to efforts to destabilize the Iranian regime or government that can reasonably be seen as illegitimate (especially those that entail violence, subversion, or active support for dissident elements)
- an end to any efforts to dismember Iran, including through promoting subversion among Iranian minority populations[31]

[31] In his March 23, 2009, address at Mashhad, presumably in response to President Obama's televised Now Ruz message to Iran, Supreme Leader Ali Khamenei made specific reference to this concern:

> "'The first measure taken by the United States was to provoke the scattered opposition groups of the Islamic Republic, and to support terrorism and disintegration in the country. They started this right from the beginning. In [m]any parts of the country, where there were grounds for disintegration, the United States had a hand, we noticed their money, and at times their agents. This cost our people much. Unfortunately, this continues. The bandits in the Iran-Pakistan border areas, we know that some of them . . . are in touch with Americans.'" (Quoted in Cole, 2009)

- recognition of Iran's right to a peaceful nuclear energy program (a right that has already been acknowledged)
- recognition of Iran's major-country status in the Persian Gulf, within the limits of other nations' own legitimate interests
- a role in the future of Iraq sufficient to reduce to an adequate degree the risk of any continuing conflict in Iraq spilling over to Iran
- a role in the future of Afghanistan to reduce the risk of insecurity stemming from the Taliban, Al Qaeda, or the trade in drugs
- respect as a country, society, and people, with both equal rights and equal obligations within the region.[32]

Neither list presented here is likely to be exhaustive. Some of the requirements likely to be posed by the United States and other countries appear, at least on the surface, to be incompatible with some of the requirements likely to be posed by Iran, and all would be subject to clarification and negotiation.[33] There is also the continuing factor that a number of regional states are in historic competition with Iran and would not want to see it readmitted to normal status under any circumstances. And, as noted earlier, there is one clear Iranian preference in regard to a regional security structure that will not be accommodated: the exclusion of the United States and other Western powers from playing a role. Indeed, for other countries in the region, and especially the Arab states of the Persian Gulf, having the United States involved—indeed, serving as the ultimate guarantor of regional security—would very likely be a requirement for their participation in such a structure, at least until the structure had clearly proved, over a lengthy period, its enduring worth. For Iran, this requirement may not, in the end, be a deal-breaker; but, as with so much else in its relations with the outside world and, especially, the United States, nothing can be known for certain before it

[32] During any negotiating process—although not necessarily as a proposal of a realistic set of goals—Iran is likely also to seek transformation of some aspects of what could be called the international order. Indeed, it could be that Iranian aspirations in this regard are designed for the purpose of pretending to be in the big leagues or, at least, of finding a way of saving face as it enters into negotiations in which, in fact, it would be the less-powerful party. Notably, see the set of proposals that Iran gave to Germany and the five permanent members of the UN Security Council in September 2009, which are very much in this vein. See Islamic Republic of Iran, 2009.

[33] Thus, Iran would almost certainly continue to demand that, in return for its own cooperation in the nuclear area, Israel's nuclear-weapon capabilities be subject to some restraint. It would further insist that all efforts to undermine its current regime cease, and it would be highly skeptical of any security guarantees offered by the United States and others, or at least of guarantees that did not include tangible and credible implementation steps. This is to say nothing about Iran's current insistence that the presence of the United States and other Western forces in the Persian Gulf is illegitimate and unacceptable. Assuming that the United States did not gain its preferred goal of zero Iranian uranium enrichment, the United States would still insist that Iran's enrichment of uranium be capped (and fully inspected) at a level well below the potential for a breakout nuclear capability and that similar limitations be imposed on plutonium reprocessing. It would further insist that external efforts to promote the democratization of Iran continue and that Iran abandon all efforts to export its revolution or to undercut peace and security efforts anywhere in the region.

is explored. Indeed, given the uncertainties in the region, which are likely to increase rather than decrease, Iran may still be led to see that it is more likely to be able to fulfill its most-basic security requirements by being "in" rather than "out," with all the consequences that the latter would imply. *Furthermore, it would not be true that, if Iran were prepared to play a positive role (as defined by others) in regional security, no formal security structure would be needed.* Such an assumption would beggar the lessons of history and the long record of misunderstandings that have, even by accident, produced conflicts.

Thus, the process and the institutions that would be set up as part of a new security structure—beginning with those proposed earlier in regard to Iraq's future—would be critical. Further, criteria regarding a hierarchy of behaviors would need to be established, commonly understood, and refined over time, supplying, in effect, formal means for determining which negative behaviors could become deal-breakers and which could be tolerated. This process would need to include a means, probably institutionalized, of determining and adjudicating violations of either explicit or implicit understandings.[34] In one partial precedent, U.S.-Soviet arms-control agreements, the political process of negotiating the agreements was often as important as the agreements themselves. In fact, in terms of building the relationship that eventually facilitated the end of the Cold War, this political process was even more important, since it proved to be an indispensable element of the basic transformation of that decades-long conflict.

However Iran chooses to proceed (i.e., whether cooperatively or not), it should be obvious that, if it did take part, all other potential participants in a regional security system or structure would have to have the capacity to conduct direct talks, leading to negotiations, with Tehran. With the onset of the Obama administration, this is now the policy of Washington. Also, at some point, diplomatic relations would have to exist among all the parties.[35] In terms of working to create a new security structure, engaging Iran would probably best proceed on a step-by-step basis, especially vis-à-vis engagement between Tehran and Washington, the relationship most fraught with impediments to negotiation. This U.S.-Iranian engagement could begin with the selection of a few areas where the two states already share common or at least compatible interests—as had long been clear in regard to Afghanistan and Al Qaeda—and where there may be other compatible interests, such as Iraq's future. However, whatever tactical steps are chosen as a starting point, success cannot be achieved unless the relationship between Iran and the outside world (especially the West and, even more particularly, the United States) is viewed in its totality. Picking and choosing what areas to deal with and in what order can work when two parties to a process are in substantial agreement about

---

[34] See Chapter Ten's discussion of the U.S.-Soviet Standing Consultative Commission.

[35] The one potential exception at the outset would be Israel. However, eventually including Israel in a regionwide security structure, or at least taking full account of its legitimate security interests, would be important for the structure's chances of long-term success.

the desirability of moving in that direction or when they already have a basically positive relationship. But it cannot work when either side begins cherry-picking issues. The former situation can build trust; the latter tends only to perpetuate discord.

Starting in this way—*with a holistic approach conducted step by step*—does not require both countries to put the past behind them in one sudden leap, but it does help to create a basis for doing business in areas where there is a good chance of agreement and, hence, for building a relationship that can be expanded over time at a pace that is mutually acceptable. A limited number of areas where interests overlap and a productive relationship can be established may remain. The issue for the creation of a new security structure for the region in which Iran plays a part is the point at which U.S.-Iranian relations (and Iranian relations with regional countries) reach a critical mass that would justify Iran's being included in, instead of excluded from, whatever formal and informal arrangements are developed. Flexibility is likely to be a key requirement in the process of creating viable security arrangements for the region from which all key players can gain sufficient benefit to make membership more attractive than "going it alone" in seeking to provide for their own security.

Asymmetric Threats

Even if all states potentially party to a new security structure for the region of the Persian Gulf agree to the concept (each calculating that playing by the rules of this new structure is more beneficial than standing apart from it), they will not be alone in determining whether such a structure is successful. Indeed, some of the most-significant players in terms of regional security are not states but nonstate actors. The term *nonstate actors* applies most consequentially (in this work) to terrorist groups, such as Al Qaeda, its affiliates, and its fellow travelers, but it also applies to any other group that either challenges the authority of a government in the region (and is prepared to do so through extralegal means) or will use violent means to prevent the emergence of a viable security structure that would stifle its ambitions. This work does not attempt to treat the full range of such movements, organizations, and activities, and it especially does not deal with those that are directed at single countries as opposed to the region or to an overall structure of security. Thus, it does not deal with either the Muslim Brotherhood in Egypt or the Mujahedin-e-Khalq (MEK) that operates against Iran. However, to the extent that the MEK and similar groups are supported by governments, their role is very much a factor that has to be taken into account, and external support for their activities would have to cease if, in this case, Iran were to buy in to a broader security structure.

Countries seeking to create a new security structure must calculate whether creating a structure that involves country X is more important than trying to overthrow the government of X or trying to promote its transformation into something more congenial. Most importantly, are U.S. efforts to promote groups dedicated to the fall of the Islamist regime in Iran through violent means more important to Washington than trying to bring Iran into a new security structure? Further, has Washington calculated, with any degree of accuracy, that it is better to work for a change of regime in Tehran in the expectation that a replacement would be more likely to be a positive player in a regionwide security structure?[36] Indeed, the U.S. government's calculation may be that the best means of promoting U.S. interests in the region is to continue trying to change the Iranian regime, the idea being either that such change can come about in the relatively near future or that a regionwide security structure that excludes Iran (and perhaps thus motivates it to play a spoiler's role) can be brought into being sufficient to meet the United States' most-basic objectives.

This issue relates very much to Iran's relationships with Hezbollah and, to a lesser and less-consequential degree, with Hamas. But it also relates directly to a major phenomenon that has emerged to a significant degree in recent years: asymmetric warfare within the Greater Middle East and Southwest Asia.[37] Of course, the kind of asymmetric warfare that the United States is now facing is not entirely new: The United States experienced it in Vietnam in the sense that a major goal of the enemy—which it achieved—was to affect popular and political opinion in the United States. The United States has also faced asymmetric warfare on other occasions, of which the most-prominent early example is the Philippine-American War of 1899–1902 (also known as the Philippine Insurrection), which, by some definitions, lasted until 1913. But a modern form of asymmetric warfare against the United States and, indeed, the West in general has become a much-more-prevalent phenomenon in recent years for three reasons. First, it is a relatively low-cost tool with a potentially disproportionate payoff against high-performance weaponry and associated elements (e.g., C4ISR[38]): In essence, it is a limited form of force equalizer. Second, it targets the politics of states (and regions) in which public opinion plays a significant role in national decisionmaking. Efforts to appeal to hearts and minds in Vietnam and in other venues during the Cold War were relatively minor facets of the overall strategy of the United States and the West, but they were a much more-important facet of strategies adopted by so-called

[36] After the June 2009 Iranian elections and the turmoil that followed, the United States faced a choice: try to deal with the regime that retained power in Tehran or, rather, wait until some outcome of the internal struggle produced a regime potentially more amenable to U.S. interests. At time of writing, the outcome of deliberations in the U.S. government on this point was not clear.

[37] The concept of asymmetric warfare as described here relates only to what insurgent or terrorist groups do (1) to regional actors (i.e., governments) and their outside supporters, such as the United States, to try to topple the government or (2) simply to oppose the presence of certain outsiders. The term *asymmetric warfare* actually has a much broader meaning. Indeed, a principal goal of virtually all military activity is to try exploiting asymmetric capabilities or techniques against the enemy in all circumstances, except wars of attrition. In fact, even in wars of attrition, one side is gambling that it can politically and economically sustain the war and the consequent heavy loss of life and property longer than can the other side.

[38] The abbreviation of *command, control, communications, computers, intelligence, surveillance, and reconnaissance.*

national liberation movements, whether pro-communist, pro-Soviet, or totally independent of others' ambitions.[39] Finally, the type of contemporary asymmetric warfare discussed in this work also targets domestic opinion in the United States (and other Western countries).

Today, in the Persian Gulf (as well as in Afghanistan and Pakistan), efforts to appeal to hearts and minds are a major element of the strategies of all the parties in conflict. These efforts have both a positive and a negative aspect: The positive aspect is that each side its trying to gain allegiances; the negative is that insurgents and terrorists are trying either to cause fear in populations that support the government or, simply, to disrupt the capacity of societies, governments, and economies to function, thus opening the way for recruiting new converts (to oppose a sitting government, for example).

Asymmetric warfare, as practiced by insurgent or terrorist groups, has three basic goals. First is the appeal to hearts and minds through either positive or negative tactics. Second is the use of relatively low-cost instruments for tactical gains against relatively high-cost instruments or classic military tactics that depend on the deployment and use of traditional force structures rather than on more-flexible approaches to warfare (e.g., counterinsurgency operations). This second aspect of asymmetric warfare is founded on an economic cost-benefit analysis of how to make counterinsurgency (for example) so costly in terms of instruments required, compared with the relatively cheap instruments used by the insurgents, that external states supporting the counterinsurgency decide that the economic and human cost of the conflict is not worthwhile. A good example is the relatively inexpensive improvised explosive device, which can destroy a high-cost armored vehicle. Of course, the economic capacities of the two sides are likely to be very different—hence, the use of asymmetric-warfare techniques by insurgents and terrorists—with the calculus of cost versus benefit thus also being very different. These calculations of relative capability and capacity for imposing *material* costs are more likely to affect local governments, especially those without major support from outside, than relatively wealthy external states that are engaged in the conflict. Calculations of *human* costs, by contrast, are likely to do the opposite, with the local government—whose very survival may be at stake—more willing to sustain casualties than the government of an outside country would be. At heart, the insurgent or terrorist group is seeking to create *force multipliers*, or, more accurately, *strategic-goal-achievement multipliers.*

The third and related aspect of asymmetric-warfare techniques is the effort to cause political change on the part of the government under attack—ideally, its overthrow—or to change the politics of its external supporters. For example, the goal of practitioners of asymmetric warfare against a local government tends to be either to

[39] Cuban President Fidel Castro's so-called barefoot doctors in Central America and the U.S.-supported contras in Nicaragua are cases in point, although they reflected opposing sides of the use of these asymmetric-warfare techniques.

demonstrate that counterinsurgency cannot succeed and that negotiations and power sharing are the only viable course or, indeed, to create so much destabilization that the government falls. The Cuban Revolution falls into the latter category. Reported Afghan government negotiations with the Taliban in late 2008 and early 2009, under Saudi auspices, fall into the former category,[40] as did Afghan President Hamid Karzai's offer in December 2009 to talk with Taliban leader Mullah Omar[41] and indications from the U.S. government that the Taliban would be part of an eventual political settlement in Afghanistan.[42]

Also important is the potential impact of asymmetric-warfare techniques on the population and politics of external states that are supporting the government and engaged in counterinsurgency or counterterrorist activities. The most-famous example of this technique's use against the United States is the Vietnam War, and the most-famous event is the Tet Offensive of 1968. The United States prevailed militarily during Tet, but the Viet Cong and North Vietnam prevailed politically (in the United States) and, hence, strategically. A more-recent example is Somalia in October 1993, when the death of 18 Americans led the U.S. government to withdraw. This—whether intended or merely accidental—was the strategy *in fact* of the insurgents, and it worked.[43]

In the broader region of the Persian Gulf, this technique has been practiced by various nonstate actors. Prominently, is was used in Iraq after the 2003 invasion, as

---

[40] See for instance, Burke, 2009; Gopal, 2008; Lamb, 2009.

[41] Gannon, 2009.

[42] See, for instance, this portion of a January 19, 2010, interview with ISAF commander GEN Stanley A. McChrystal:

> FT: Do you think then that it would be conceivable that this conflict could end with senior figures in the Taliban perhaps playing a role in a future government in Kabul?
>
> Gen McChrystal: It's hard to speculate about individuals, but I think that anybody who dedicates themselves to the future and not the past, and anybody whose future is focused on the right kinds of things for Afghanistan, under a constitutional fair umbrella, then I think it's likely that it will be a wide participation.
>
> FT: So it sounds from what you're saying that you wouldn't be on principle opposed to the idea that some of the insurgent leaders that you are fighting now might one day be part of a future administration in Kabul?
>
> Gen McChrystal: It wouldn't be mind [sic] to choose, it would be the Afghan people's decision to choose. But I personally believe that the Afghan people will want to represent themselves from the entire Afghan population. (M. Green, 2010)

[43] For the United States, a rule of thumb is that the American people will support the investment of blood and treasure in a foreign conflict that does not clearly and directly impinge on the security of the U.S. homeland only if doing so meets three criteria: it is in the nation's interests, it comports with American values, and there is a strategy for winning or gaining some other clearly valuable outcome (sometimes called an *exit strategy*). This calculus—in particular, that American interests are clearly at stake—did not support a case for U.S. engagement in efforts to stop the war in Bosnia (1995) and the ethnic cleansing of Kosovo (1999). In both instances, the United States thus relied solely on air power in the hope of minimizing U.S. casualties. Other NATO allies made similar calculations. In the event, neither bombing campaign led to a single combat-related U.S. or allied death. Somalia is another good example of a conflict in which American values were in play but the United States' interests were not—hence, U.S. forces were withdrawn after sustaining a relatively small number of casualties.

insurgents, terrorists, and other opponents of the government in Baghdad tried to cause a level of U.S. casualties beyond that which the American people would believe to be worthwhile given U.S. interests and values. This use of an asymmetric-warfare technique to try affecting U.S. domestic opinion and, hence, popular support for the war has not been a coordinated strategy by the different groups that have sought to gain power in Iraq and, in the process, to oppose the United States, but it has had a significant impact on the willingness of the American people to continue being engaged militarily. It also led to a reassessment in the United States of the importance of Iraq in terms of what goals are worth fighting for. Indeed, the United States' decision to draw down its forces in Iraq, a decision ratified by the Security Agreement, which calls for the departure of U.S. forces by the end of 2011, was thus a *timetable-determined decision*, not a *conditions-based decision*. There has been virtually no discussion in the United States about what it will do if the drawdown is accompanied by a large-scale increase in violence in Iraq and by, perhaps, the fall of the government and, even, the rise of civil war.[44] The American people and, now, their government have essentially written off Iraq as a place where it is worth spending much more blood and treasure. Despite the success of the so-called surge, asymmetric warfare in Iraq against the United States and its Coalition partners has, in effect, succeeded, if only on a slow-roll basis.[45] Of course, the United States' bet that it will leave behind an Iraq that can more or less take care of itself in terms of promoting internal stability may prove wise. Indeed, the process of the drawdown is designed, at least in part, to make the case that the United States is leaving because the job is done and not because it has lost the will to persevere—a not-inconsequential difference.

---

[44] It was within this context that an official related the following remarks reportedly made by Vice President Biden during a visit to Baghdad in July 2009:

> If "Iraq were to revert to sectarian violence or engage in ethnic violence, then that's not something that would make it likely that we would remain engaged because, one, the American people would have no interest in doing that, and, as he put it, neither would he nor the president," the official said. (Bakri, 2009)

[45] The essence of the debate that took place in the United States after the initial phase of the war in Iraq—following the dearth of free debate in the United States before the invasion—focused both on the proximate reasons for the war (notably, whether Iraq had or was developing WMDs) and on the United States' interests in invading. In strategic terms, the drawdown of U.S. forces in Iraqi is possible because the debate over what the United States did in 2003 has been essentially settled: There was no compelling reason for the invasion. Ironically, however, as argued here, the invasion created its own reasons for U.S. involvement because Iraq was taken out of the balance of power in the region (including against Iran) and because it became a locus both for instability and for the potential projection of insurgency or terrorism into other countries. Whatever the accuracy of strategic calculations, politics in the United States have decreed that a continued U.S. force presence is not required to achieve U.S. interests. The surge, in effect, bought the United States *breathing space*, to borrow a term of art used during the Vietnam War (another Vietnam-era term of art that could be applicable was the *decent interval* allowed before withdrawing forces and facing the potentially negative consequences). As argued here, the test will come when U.S. forces have been drawn down significantly and violence either increases or does not. Thus, like the invasion, the timetable-determined drawdown was conceived with a "hope and a prayer," although, in the case of the drawdown, there may be a more-solid basis for this approach to strategic calculation.

This work has advanced two basic goals for Iraq regarding long-term U.S. interests in the Greater Middle East.[46] However, neither emphasizes, at least in the short-to-medium term, the democratization of Iraq, which was one of the principal goals of U.S. engagement in earlier years. That goal has now, in practice if not in policy, been put off.

For their part, an effort made by the Taliban and Al Qaeda, coordinated or not, has aimed to convince the people and, hence, the politics of external countries involved militarily in Afghanistan that they are paying too high a price (in this case, primarily in terms of casualties) to justify continued engagement. This tactic has only just begun to affect the United States—after all, the attacks of 9/11 were directly connected to Al Qaeda and the Taliban regime in Afghanistan—but has, for some time, been having a profound impact on many of the United States' NATO allies that have forces in the country. For several of the allies, a relatively small loss of troops or even the perception that they are likely to suffer even a small number of casualties has played a major role in the imposition of so-called caveats, which limit where forces can be deployed and what they can do, and has led to domestic political debates in several allied countries about the possibility of withdrawing those forces that have been deployed. How this will play out is uncertain. At the time of writing, these domestic pressures continued to mount in some allied countries, although some other countries have acceded to President Obama's appeal for more NATO forces to be sent to Afghanistan. Pledges to do so had, by mid-January 2010, totaled about 9,000 troops.[47]

The issue of asymmetric warfare is relevant to a new security structure in the Persian Gulf in at least three ways. First is the question whether all participants (states, of necessity) will be willing (1) to foreswear using asymmetric-warfare techniques against each another in pursuit of their own political, ideological, or other objectives and (2) to oppose the use of these techniques by others, including nonstate actors. Iran is the key focus of this question when considering states, but it is not alone. Indeed, in terms of the flow of moneys from a regional state to groups that, in one form or another, help to promote Islamist extremism and terrorism, Saudi Arabia has to be ranked number one. (Note that such support is not an act of the government per se but rather of wealthy individuals and various religious groups.) Clearly, if there is to be any hope for a new security structure in the region of the Persian Gulf, the Saudi government and the governments of all the other Sunni Arab states have to put a full stop to such activities, beginning in Iraq.

In addition, each country that wishes to take part in a regionwide security structure must calculate the relative advantage of being so engaged compared with supporting elements in Iraq that are not just contending for power—an inevitable activity—but are also using violent means to achieve their ends, which is a form of asymmetric

[46] See Chapter Two.

[47] Rasmussen, 2009c.

warfare. Again, there may be levels of tolerance for such behavior within a viable security structure. It could even be argued that, if Iraq can be stabilized or if it otherwise poses little or no threat to broader Persian Gulf security objectives, then outside support for elements contending for power, even with violent means, might not matter overmuch. But this is a doubtful proposition. In at least one case, it is essential that support for one group that practices asymmetric warfare—the PKK, operating from Iraqi Kurdistan against Turkey—be ruled illegitimate and unacceptable and that its efforts be opposed by all countries that value a new security structure, within which Turkey has to be a partner for the structure both to be encompassing and to have a chance of achieving long-term success. The same would need to be true of the MEK in regard to Iran.

A second way in which asymmetric warfare is relevant to the region of the Persian Gulf and the prospects for a new security structure is that there needs to be broad-scale agreement to *oppose* any use of asymmetric warfare. Of course, achieving this agreement would be hard unless each and every instance in which such asymmetric warfare might be practiced were considered. Two cases that are not in the Persian Gulf stand out. One relates to Lebanon and its evolution as a viable, unitary state. If there is to be general agreement to oppose asymmetric warfare, then its practice has to cease in Lebanon, and this will have political implications for internal groups and external states. (It may be possible to draw a distinction between internal groups and external sponsors: To be considered valid members of a regional security structure, states would need to join the understanding proposed here and adhere to its tenets.) The other case is the Arab-Israeli conflict and, most particularly, relations between the Israelis and the Palestinians and, even more particularly, Israel and Hamas. Although it might be desirable to try getting everyone involved in this conflict to forswear asymmetric-warfare techniques, this will remain impossible as long as there is no resolution of deeper political issues and as long as there is a vast disproportion between the two entities' conventional military power.

A third aspect to the impact of asymmetric warfare in the region of the Persian Gulf is that efforts by local states to restrain it, oppose it, and refrain from either sponsoring it or condoning it are *required* if these states are to expect the sustained involvement of external actors, especially the United States and European countries. Indeed, as each country in the region calculates whether its own security will benefit from continued U.S. or other Western engagement, whether that country pursues that engagement on its own or through a formal security structure, it will have to decide whether it will act to reduce the risks to Western military and civilian personnel who are deployed in the region on its behalf. Countries in the region cannot turn a blind eye to asymmetric warfare in their midst and expect the United States and others to be engaged in promoting security and stability in a capacity that extends beyond any efforts that these external states would consider to be so important to their own interests that they would be willing to sustain significant costs, including human costs.

Regional Reassurance

As argued earlier, a principal reason for seeking to develop a new security *structure* for the Persian Gulf as a key element of its operative security *system* is to try finding a means whereby the United States can achieve its strategic and security goals in the region at less material and human risk and cost than at present and than would otherwise be required if the current efforts of the United States and others to promote regional security were simply continued without major change. A subsidiary reason for this effort is the possibility, indeed, the likelihood, that the American people will not want indefinitely to continue the current level, or even a reduced level, of engagement of the size, character, and quality of that required by current policies and approaches, including the recommitment of the Obama administration to sustaining sizeable levels of U.S. military forces in Afghanistan—first, by increasing them by 21,000[48] and then, in December 2009, following lengthy debate,[49] deciding to send an additional 30,000 U.S. military forces.[50] A third reason for this effort is the *opportunity costs* that the United States—and, potentially, allies and nonregional partners also involved in trying to create a reasonably stable security situation in the region of the Persian Gulf and environs (extending as far as Afghanistan and Pakistan)—would otherwise incur. These opportunity costs relate not just to alternative uses of economic resources or even to alternative uses of military forces but also to the time and attention that is currently devoted to security issues in the Greater Middle East, rather than to East and South Asia, in particular, and, most notably, to the emergence of China and India as great powers. There is also the issue of the gradual reemergence of Russia as a great power, some of whose security-related interests—especially those in the regions of "privileged interests," as asserted by Russian President Dmitry Medvedev[51]—are already proving to be at odds with those of the United States, its allies, and its partners.

The development of a new security structure cannot just be an excuse for a reduction of U.S. involvement in the region, however. Indeed, any alternative to current policies, including patterns of deployment of U.S. military forces and nonmilitary assets to the region, must also account for expectations on the part of regional states about the future role of the United States in and with regard to the region, whether the United States works on its own or in league with European and other allies and partners, such as those in Asia (notably, Australia). Furthermore, expectations are not confined to

---

48 The White House, Office of the Press Secretary, 2009.

49 The debate about the appropriate level of U.S. and other troops in Afghanistan had as a central feature analysis made by the ISAF and U.S. force commander in Afghanistan, General McChrystal. A declassified version of the report was published by the *Washington Post* on September 21, 2009. See "COMISAF Initial Assessment (Unclassified)—Searchable Document," 2009.

50 Obama, 2009e.

51 See P. Reynolds, 2008.

countries that look to the United States to protect, whether actively or latently, their security and other interests:[52] Countries that see the United States as enemy or competitor—e.g., Iran in its current state—also have expectations. Put simply, there is no country other than the United States that, given current configurations of military and other forms of power and given current configurations either of leadership or of potential for political commitment, can play a major security role in the Persian Gulf region. Relevant security matters include those related to the future of Iraq; to the balancing or managing of Iran's role in the region; to the possible need for decisive action in the event of threats to freedom of the seas (e.g., threats to shipping that transits the Strait of Hormuz); to the countering of terrorism; to potential tensions, strife, or even conflict involving regional states that could have broader implications (e.g., for the flow of oil); and to crossborder military or subversive activities (e.g., PKK attacks on Turkey and significant Turkish retaliation or tensions and strife involving Yemen) that would be of sufficient magnitude as to pose a risk either of escalation or of broader destabilization in the region.

This means that, at least for the foreseeable future, a new security structure for the Persian Gulf would be unlikely to allow the United States to divest itself of all the tasks that either it has taken upon itself or are imputed to it on a contingency basis by local countries, nonstate actors, or nonregional countries (e.g., European or Asian allies and such countries as Russia, China, and India). At the same time, there would still be value for the United States and, arguably, for other parties in seeking alternatives to the United States' having to play as preeminent a role as it does now, including the need for it to provide the vast bulk of the external military resources and assume the largest share of risk in terms of potential casualties in conflict.

If such a reduction in U.S. responsibilities does prove possible, it would entail a number of requirements that are discussed throughout this work. One that is both hard to define and hard to translate into concrete terms but is of major importance is the perception, both in the region and more broadly, that the United States is in the region to stay in ways that will continue to make it, in effect, a permanent power in the Greater Middle East. Being able to foster this perception at lower cost and risk while still protecting its interests, including the retention of substantial political influence, would be ideal from the United States' perspective. Such a relative reduction of

---

[52] According to J. Russell, 2007,

> GCC states are also dismayed that the Iraqi bulwark against Shia Iran has been removed by the toppling of Saddam Hussein's predominantly Sunni Baath regime in Iraq, and the civil war and growing Iranian influence in Iraq that have followed this regime change. They have been concerned that Sunni Arabs have been marginalized in the new Iraq, that Iran exercises too much influence over the Shia Arab parties that dominate the new government there, that Iranians in Iraq may engage in subversive activities against them, and that the civil war may actually spill over into their own states. GCC states are uneasy that Iranian influence is growing in a "Shia crescent" across the region, particularly in the Levant, and particularly because of the unresolved Arab-Israeli conflicts.

responsibilities would be in the U.S. interest, of course, *unless the United States were to seek to be the dominant power in the region for its own sake* and, thus, also be willing to incur the lion's share of cost, risk, and responsibility—perhaps not a very good bargain. But, as it reshapes its policies and postures toward the region, the United States must provide reassurances that it will indeed not abandon its commitment to the region's security, and this will more easily be said than done.

Demonstrating a long-range commitment is not just about force deployments, the acquisition of basing rights and the practical use of them, and the practice of associated diplomacy and engagement of private-sector entities. It is also a matter of demonstrating that the domestic politics of the United States will permit it to sustain its involvement in the region such that all who are engaged in the region or affected by it conclude that the United States will be a permanent regional power, ready and willing to defend its own interests and the relevant interests of other external powers and regional actors, both existentially (as a perceived inclination to be engaged and to act when needed) and as formally and clearly defined for all to see.[53] The task, therefore, is to devise a pattern of direct U.S. involvement in the region, both military and civilian and both onsite and over-the-horizon, such that the U.S. administration, the U.S. Congress, and the American people will be prepared to sustain such engagements for the indefinite future and be seen to do so.

[53] President Obama announced in December 2009 that, "taken together, these additional American and international troops will allow us . . . to begin the transfer of our forces out of Afghanistan in July of 2011." This was a "time-based" rather than "conditions-based" judgment, even though the president added that "[the United States] will execute this transition responsibly, taking into account conditions on the ground" (Obama, 2009e). This telegraphing of the time at which U.S. forces will begin to leave Afghanistan could erode the confidence of some observers in U.S. staying power. This development will depend, in part, on how much has been achieved in Afghanistan by that time, on what else the United States is doing, and on exogenous events (e.g., a major change in U.S. relations with Iran). In addition to efforts related directly to Afghanistan and Pakistan, the United States could also profit by beginning the process of creating a security structure for the Persian Gulf.

CHAPTER FIVE

Elements of Security Reassurance

Five key elements of U.S. activity within the Persian Gulf region that relate to fulfilling the requirement for security reassurance are as follows: the withdrawal of U.S. forces from Iraq, U.S. policy and approaches regarding Iran, U.S. forces in and near the region, formal security guarantees, and a U.S. nuclear guarantee. These and other elements of security reassurance are discussed in this chapter.

The Withdrawal of U.S. Forces from Iraq

The United States should conduct the withdrawal and repositioning of its forces from Iraq in a way and at a pace such that it is seen to be acting within a valid strategic framework regarding its own interests and those of key regional and extraregional countries rather than as the result of war weariness. This can help to offset the potential impact in Iraq of the timetable-based withdrawal, which can be exploited by opponents of a stable Iraq who try to wait out the U.S. departure. Creating a valid strategic framework includes calculating requirements in relation to interests; analyzing costs versus benefits; reassessing the importance of different objectives (particularly the value of full political, ethnic, and religious reconciliation in Iraq or the development of democracy at an advanced level); and considering alternative actions and activities, including the steps listed in the Iraq discussion in Chapter Four, and the development of a viable security structure for the region as a whole.[1]

U.S. Policy and Approaches Regarding Iran

The United States should review its policies and approaches toward Iran and Iran's roles within the region and beyond such that the United States will be seen as advancing its own interests, both narrowly and broadly defined, and as protecting the basic

[1] The Obama administration's adoption of a *timetable-based* rather than *conditions-based* Iraq withdrawal plan could prove to be at variance with the development of a valid security framework.

security interests of other regional countries. This does not mean that the United States should necessarily adopt the views of other states as its own. Indeed, one of the tasks involved in trying to create a new regional security structure is to provide incentives for Iran to play a positive rather than negative role and to obtain Iran's reassurances about its ambitions toward other regional states or the region as a whole. If Iran were to follow this positive course—and, in terms of political reality, this must include stopping short of attaining the capacity to build nuclear weapons (or, if it proved to be too late to keep Iran from reaching that point, Iran would have to accept an inspection regime of the utmost intrusiveness)—there would need to be recognition by these other regional states that Iran was being cooperative. Gaining Iran's acquiescence to such a course—including security assurances from the United States and others in exchange for Iran's agreement to act in ways that demonstrate a nonhostile approach to its neighborhood—could in fact prove easier than getting some of the Arab littoral states of the Persian Gulf to appreciate such behavior and to moderate their own apprehensions or general desire to see Iran cut down to size. Such is the traditional lot of small countries in the vicinity of larger powers. But this is one reason for the United States to demonstrate that it both can and will remain a regional power. Of course, the United States has to resist efforts by local states to blow any Iranian threat or challenge out of proportion in an attempt to gain an unneeded level of insurance from the United States that could in fact seem as provocative to Iran, a difficult but important balance to strike. From the United States' point of view, certainly, that course would be self-defeating and would retard whatever positive developments there could otherwise be in Iranian behavior. In addition, as the United States develops new policies toward Iran, these need to be based on the principle of mutual respect, a cardinal requirement for both countries and societies.

Bolstering Regional Defenses

One obvious way to demonstrate U.S. commitment to the security of states in the Persian Gulf is to bolster individual national defenses, including through arms sales, whether by U.S. firms or by those of other Western states. This policy has a long history and will no doubt continue, whether driven by demand or supply, in response to perceived security needs on the part of regional states or reflecting Western efforts both to shore up these states' sense of security and to earn revenues from arms sales.[2] As concerns among both Gulf Arab states and the United States about Iran's development of ballistic missiles, its nuclear program, and its intentions in the region have increased, so have the supply of arms to the Arab states and ancillary efforts by the United States

[2] Arms supply to Persian Gulf countries and its potential impact on regional stability has been a topic of consideration for at least decades. See Kennedy, 1975.

to bolster these states' defenses. These efforts have included the provision of antimissile systems to four regional countries,[3] coordination of air defenses, joint exercises, and U.S. arms sales to regional countries totaling more than $25 billion since 2008.[4] There are other factors to be considered, however. One is whether the regional states will be able, anytime soon, to turn armaments into effective military forces. Another is whether this rapid increase in military supplies will do more to deter potential Iranian military action (or exploitation of a possible nuclear-weapon capability in the future) than to intensify a conventional arms race with Iran that could be uncontrollable.[5] A third factor is whether Persian Gulf Arab states' expanded military capabilities will prove useful in the absence of a regional security structure such as the one being considered in this work. A fourth is whether such a buildup would tend to foreclose the possibility of including Iran, at some point in the future, in such a structure, which would, in effect, further lock in confrontation between the Gulf Arab states and Iran.

U.S. Forces in and near the Region

The United States could reposition military forces in and in relation to the region such that there would not be a widespread perception in the United States that the size and character of its deployments were imposing risks and costs (human and material) that outstrip the intrinsic worth to the United States of any key element of its regional policies and engagements. The repositioning would have to take into account the potential risks associated with military deployments and some civilian deployments, which could have a lightning-rod effect on Islamic extremists or local nationalists who could increase the risks to U.S. personnel or even heighten the chances of conflict well beyond the level the American people would tolerate. These forces must also not be deployed in types, quantities, and configurations that could prove destabilizing. Such calculations need to be as exact as possible, and there would be great value in communicating to all parties what the United States was doing and in consulting widely to avoid misunderstandings.[6] Such U.S. deployments would need to focus either on regional countries where the sheer presence of U.S. forces would not be likely to have a lightning-rod

[3] See Sanger and Schmitt, 2010.

[4] See Warrick, 2010.

[5] See Chapter Ten on arms control and CBMs.

[6] This issue has already arisen in relation to U.S. force deployments in the Persian Gulf and was evident in a scuffle between U.S. and Iranian naval units in January 2008, discussed briefly in Chapter Ten. Force presence can convey commitment and, properly done, can demonstrate the credibility of commitments. But it can also appear to another party to be confrontational, and it can become destabilizing—communicating by *fact* as opposed to *words* and potentially operating in opposition to the words—*a distinction classically known as action policy* versus *declaratory policy*. Hence, the value in direct contacts. As discussed elsewhere, this failure to understand the power of action policy in seeming to communicate intention was a cardinal error committed by Egypt

effect, as in Qatar and Bahrain at present (but this is by no means assured for the future); where even the presence of U.S. forces in the immediate region would not be seen to have negative effects; or at some distance, where they would truly be out of sight but not so far away that they would be truly out of mind. The U.S. military base on the British island of Diego Garcia would certainly be out of sight; uncertain is whether U.S. forces based there would be so out of mind—i.e., irrelevant for security purposes—that the Persian Gulf states concerned would not have confidence that U.S. forces could be reinserted in a timely fashion.

Of course, a balance has to be struck. To provide reassurance to local countries (or to deter other countries), U.S. force deployments would also have to be sufficiently close by to be capable of being inserted into crisis or conflict situations when needed, and they would have to be sufficiently potent—and seen to be so—to have a critical impact. The balance can be struck, in part, whether or not there is a significant withdrawal of U.S. forces from the immediate region, through several means: holding exercises,[7] creating partnerships with regional militaries (when doing so would not become a source of internal instability), training local forces, and undertaking other activities that could help to shape, within individual countries, conditions that are more likely to foster a benign, rather than malign or hostile, environment for an external Western presence and involvement.[8]

Furthermore, it is also important to recognize that the "threats" that some Arab states of the Persian Gulf see to their security derive, at least in part, either from the risk of asymmetric warfare (especially terrorism, including, now, emanating from Yemen) or from nonmilitary pressures from Iran, especially the activities of Iranian migrants,

---

when it deployed forces in the Sinai Desert in May 1967 and also demanded that the UN Emergency Force that had been deployed there depart. See Howard and Hunter, 1967.

[7] There is a long history of holding exercises in the Persian Gulf that involve externally based U.S. forces. There could also be merit in conducting exercises analogous to the Cold War–era Return of Forces to Germany (better known as REFORGER) exercises in Europe, although heed must be given to all the cautions about balancing risk and benefit in terms of local perceptions, including the unwarranted raising of anxieties about a permanent reinsertion of U.S. military power or even a U.S.-led invasion.

[8] An added consideration is the potential threat, in the future, from Iranian ballistic missiles, perhaps armed with either conventional or nuclear warheads, if Iran were, indeed, to proceed with developing these weapons. This issue will matter less to the countries of the Persian Gulf region than, in particular, to Israel, even if Iranian missiles were armed with only conventional warheads with high accuracy. Certainly, an Iranian ballistic-missile capability would raise questions throughout the region about possible U.S. responses and, hence, U.S. staying power. Increased Iranian military capabilities would also have an impact on regional perceptions of "security," writ large. This factor would, of course, be most important if Iran had nuclear weapons, but it could also be important if Iran had significant capabilities for fielding and, perhaps, employing ballistic missiles with conventional warheads (e.g., an advanced version of the Shahab-3 missile). This issue was addressed in the decision taken by the United States in September 2009 not to deploy antimissile defenses in Poland and the Czech Republic designed to counter an intercontinental ballistic missile (ICBM) threat from either North Korea or Iran and instead to pursue defenses of a "distributed" character against shorter-range Iranian missiles. This decision could, in time, have a positive impact on regional assessments of U.S. commitment to regional security against a potential Iranian missile threat. See Gates and Cartwright, 2009.

Iranian investment and trade practices, and Iran's potential manipulation of religious factors. (Iran's ability to exploit this last-named factor needs to be kept in perspective: The perception of Iran's influence on coreligionists in Arab states may be inflated compared with the reality.) Diplomatic intimidation by a nuclear-armed Iran would be particularly troubling for regional states, even though it is not entirely clear that merely possessing one or more nuclear weapons could grant Iran effective diplomatic leverage. Thus, U.S. reassurances to regional states against these potential threats or challenges, which are more likely to emerge than open military attack from, say, Iran, need to reflect the fact that military instruments, especially high-performance military instruments, may not be particularly relevant other than to show a general interest in regional security or for purely classic deterrence of a major military action, including the potential use of a WMD. For example, the United States must be able to implement a full range of responses to terrorism in the region. At the same time, military-based deterrence against subversion is unlikely to apply when instruments for making deterrence credible are far out of proportion to the provocation against which such deterrence is designed. Local countries will need to rely more on their own devices, including control of migratory and investment flows; to work with the United States and others on counterinsurgency and counterterrorism policies (if relevant); or to accept that a reduction in internal challenges from, say, Tehran, may be most likely to come about under circumstances in which the United States and other Western countries have managed to forge workable relationships with Iran.

Formal Security Guarantees

The United States could consider providing formal security guarantees to regional states against aggression from their neighbors and, potentially, from sources external to the region. However, such security guarantees could be ineffective in the case, say, of threats of terrorism or internal subversion, especially when there is no clear "return address" or when the perpetrators of violence have no tangible assets to be put at risk.

Over the years, the United States has generally been reluctant to provide such guarantees to states in the Middle East (except for Turkey in its capacity as a NATO ally). Even during the Baghdad Pact/CENTO period, the United States did not provide direct guarantees to regional member states,[9] and, despite the closeness of the security relationship between the United States and Israel and the strong political underpin-

[9] See U.S. Department of State, undated. The United States did conclude a mutual defense-assistance agreement with Pakistan in 1954, but that agreement did not contain any promise to defend Pakistan against aggression. See Mutual Defence Assistance Agreement, Pakistan and the United States of America, 1954. This agreement was reinforced in 1959, and the United States also entered into bilateral agreements of cooperation with Turkey and Iran. None of these agreements, however, called for the United States to defend any of these countries (except, as just noted, Turkey). See Khan, 1964.

nings of U.S. willingness to defend Israel against a serious attack, there is no formal U.S. commitment to Israel's defense. Other than the case of Turkey, there have been two exceptions to this practice of not providing direct guarantees. The first was the generic guarantee of regional security against "outside force" contained in the Carter Doctrine of 1980: Iran was, in fact, the implied recipient of the guarantee against possible Soviet aggression, but it was not named.[10] The second is contained in the U.S.-Iraqi Security Agreement of November 2008:

> In the event of any external or internal threat or aggression against Iraq that would violate its sovereignty, political independence, or territorial integrity, waters, airspace, its democratic system or its elected institutions, and upon request by the Government of Iraq, the Parties shall immediately initiate strategic deliberations and, as may be mutually agreed, *the United States shall take appropriate measures, including diplomatic, economic, or military measures, or any other measure, to deter such a threat.*[11]

But, for security guarantees proffered to Iraq or any other country in the region to be credible, they need to contain commitments that the American people are willing—and believably so—to honor and are defined in clear relation to the U.S. interests that would be at stake and could, conceivably, come under challenge.[12] Guarantees in the case of overt military aggression across a land or sea frontier by a neighboring (or more–far-flung) state would very likely meet those criteria, but what about subversion, state-sponsored or state-facilitated or state-tolerated terrorism (areas in which Yemen has emerged as a source of concern), or state support for a domestic insurgency? What would be the triggering point, and how would this be defined? Certainly, there would need to be formal criteria, presented in a documented form that could range from a memorandum of understanding that would clearly have the tacit blessing of the U.S. Congress to an even-more-credible formal treaty subject to U.S. Senate ratification. Any such guarantees would also have to be defined either in generic terms (e.g., applying to attacks by any country against any regional country—*tous azimuts*, as it were) or in specific terms (e.g., applying to aggression by Iran against Arab states of the Persian Gulf or to aggression by the Iraqi Kurds against Turkey under circumstances, such as terrorism, that are not already clearly covered by Article 5 of the Treaty of Washington).

---

[10] See Carter, 1980.

[11] See Agreement Between the United States of America and the Republic of Iraq on the Withdrawal of United States Forces from Iraq and the Organization of Their Activities During Their Temporary Presence in Iraq, 2008, Article 27 (emphasis added). This aspect of the agreement needs to be contrasted with Vice President Biden's comments in Baghdad in July 2009.

[12] This provision of the U.S.-Iraqi agreement attracted little notice in the United States. It is not at all clear that the United States would want to honor the agreement's provisions after its forces depart Iraq.

As of now, the most-important aspect of security in the region for the United States to consider is whether it should give security guarantees to countries against possible Iranian aggression *in the absence of a direct threat.* This issue was acknowledged by Secretary of State Hillary Clinton in July 2009, when she suggested that, if Iran acquires a nuclear weapon, the United States could extend a "defense umbrella" over the region.[13] Doing so would likely "lock in" U.S.-Iranian confrontation and make it more difficult to negotiate a change in relations. By contrast, the United States could offer guarantees to local Arab states if there were an irreparable breakdown in U.S.-Iranian diplomacy and if Iran were posturing itself such that other regional states would need to have added assurance of U.S. security engagement in the region. Of course, the fact that such formal agreements proved necessary to reassure regional countries of the credibility of U.S. support for Persian Gulf security could indicate to these countries that the United States did not have its own national security reasons for being engaged and, therefore, such guarantees might not be worth as much as might appear on the surface—a point to be pondered.

A U.S. Nuclear Guarantee

As of now, there is one set of circumstances that could raise the issue of whether the United States should consider providing nuclear guarantees (a nuclear umbrella) to regional countries: Iran's decision to proceed to develop nuclear weapons or its attainment of a breakout capacity. Historically, the United States has been chary of extending such guarantees, especially outside of formal treaty commitments. Thus, these guarantees have been limited to members of NATO and to Japan, the Republic of Korea, and Australia.[14] Even Israel does not have a formal nuclear commitment from

---

[13] "U.S. 'Will Repel Nuclear Hopefuls,'" 2009, quotes the following statement made by Secretary Clinton during an interview for Thai television:

> "If the U.S. extends a defence umbrella over the region, if we do even more to support the military capacity of those in the Gulf, it's unlikely that Iran will be any stronger or safer because they won't be able to intimidate and dominate as they apparently believe they can once they have a nuclear weapon."

[14] Although it is a member of the Australia, New Zealand, United States Security Treaty, New Zealand lost the protection of the U.S. nuclear umbrella when it declared itself a nuclear-free zone in 1984. See Ministry for Culture and Heritage, 2008. Of course, this U.S. action has had no practical effect. At the same time, the United States has given what are called *negative security assurances* regarding nuclear weapons. These assurances were first made in 1978 and were then modified and reaffirmed in 1995 in a statement issued by then–Secretary of State Warren Christopher:

> The United States reaffirms that it will not use nuclear weapons against non-nuclear-weapon States Parties to the Treaty on the Non-proliferation of Nuclear Weapons except in the case of an invasion or any other attack on the United States, its territories, its armed forces or other troops, its allies, or on a state towards which it has a security commitment, carried out or sustained by such a non-nuclear-weapon state in association or alliance with a nuclear-weapon state. . . . The United States affirms its intention to provide or support immediate assistance, in accordance with the Charter, to any non-nuclear-weapon State Party to the Treaty on the Non-

the United States, although Israel's own nuclear arsenal means that such a guarantee is not very important and, in fact, would be far less credible than Israel's capability and willingness to act in its own interest.

Thus, considering taking countries in the Persian Gulf region under the U.S. nuclear umbrella would represent a significant departure from past U.S. practice. In all likelihood, it would also be excessive, since U.S. conventional military power would be more than sufficient in fact, if not in the perception of local countries,[15] to inflict on Iran what nuclear theory refers to as *unacceptable damage* in response to Iran's use of one or more nuclear weapons. (Such Iranian use is most unlikely to occur except in response to a major military attack already taking place against Iran or in an act of simple insanity. Neither case is deterrable, in any event.)[16] The United States might choose to reinforce security pledges to regional states (an idea floated by Secretary Clinton in July 2009) and to reinforce quick-reaction air attack capabilities within the region as a visible, conventional-force deterrent. But it is not evident that a threat to retaliate with nuclear weapons against Iranian first use of its own nuclear weapons would be needed for deterrence as long as Iranian leaders made rational decisions.[17]

> Proliferation of Nuclear Weapons that is a victim of an act of, or an object of a threat of, aggression in which nuclear weapons are used. (W. Clinton and Christopher, 1995)

Note that this provision is silent about the possible use of nuclear weapons to fulfill this pledge. This pledge, agreed to also by China, France, Russia, and the UK, was reinforced in United Nations Security Council Resolution 984.

[15] This is an important distinction. But, given the complexities and risks involved in providing nuclear guarantees to other nations, including the definition of the precise circumstances in which the guarantees could be invoked, the United States would need to be careful lest the Arab states' demands for guarantees that could be excessive compared with U.S. judgments of what is really required actually make matters worse by seeming provocative to Iran and, perhaps, making a dialogue on these issues, designed to help promote regional stability, more difficult even than it is at present.

[16] In theory, there could be so-called intrawar deterrence, a situation in which Iran, while under conventional attack, might be deterred from using nuclear weapons because of the credible prospect of sustaining further punishment, perhaps including U.S. nuclear strikes. However, a country conducting a major conventional attack on Iran could not rely on such a response, and, therefore, this point of theory is not something that could be counted on in practice.

[17] Cold War experience with U.S.-Soviet nuclear confrontation demonstrates the complexities of making judgments in this area and the risks inherent in a nuclear standoff in which one or the other of the parties in confrontation lacks a second-strike capacity. The imponderables in this area are another argument for forestalling Iranian acquisition of a nuclear capability, using (diplomatic) means short of warfare. At the same time, the West needs to make clear to Iranian leaders the risks to their own security that they would run by acquiring a nuclear capability. This communication should be accompanied by willingness in the West to consider Iran's legitimate security requirements, something that does not appear to have been done.

Other Elements of U.S.-Provided Security

The United States needs to conduct its diplomacy either to foster a positive appreciation among Persian Gulf Arab states of what it is doing and is prepared to do in the future or to produce a deterrent effect in regard to potentially hostile countries. This diplomacy could include promoting the development of a regional security structure, provided that related efforts are credibly characterized not as an effort to gain political dominance in the region or as a means for the United States simply to withdraw from the region or to wash its hands of the region's problems. Rather, these efforts must be perceived as creating the basis for U.S. involvement in the region that—whether the United States is engaged as active participant, as mentor, or as residual guarantor of security—both contributes to a sense of security on the part of local countries and is sustainable in U.S. domestic politics. Indeed, the process of promoting such a regional security structure could work, over time, to create buy-in on the part of local states, other external states, and the American people. Success can breed further success. Critical aspects are transparency in developments; a "rule of reason" about what is being done; and a sharing of burdens, risk, responsibility, influence, and decisionmaking. "Made (solely) in the USA" is not a valid basis for creating a new security structure for the Persian Gulf region that can fulfill the various objectives this work establishes. Sharing *risks and responsibilities* with other external powers in a new security structure would likely require a greater degree of sharing both *decisionmaking* in regard to providing regional security, broadly defined, and *political and economic influence* in the region. It would also mean that regional countries, acting within a new security structure that included some level and appropriate character of assurance or guarantees from the United States and, perhaps, other external powers, would be expected to assume a much-larger share of responsibility for their own security, including not just what they do in terms of military activity and regional diplomacy but also what they do to foster improved relations, in general, within the region. *Processes* to reinforce this objective would be required.

CHAPTER SIX

The Arab-Israeli Conflict

The subject of this chapter is a key element of U.S. activity that is sufficiently important and has such a long and checkered pedigree that it needs to be discussed on its own: the Arab-Israeli conflict and the role that the United States either does or does not play in trying to resolve it.

All calculations about a security structure for the Persian Gulf—indeed, all calculations about the politics of the Middle East and about U.S. engagement in the region—continually return to the question of the Arab-Israeli conflict and, more particularly, to the conflict between Israel and what, for simplicity, this work refers to as Palestine. It is possible to argue that the nations of the Persian Gulf region have little or no security interest in what happens in the zone of Arab-Israeli conflict. None has anything immediately at stake, other than the future of Jerusalem and its Muslim holy places.

However, in the Muslim world, especially in Arab countries, popular opinion (commonly referred to there as the *Arab Street*) is strongly motivated by the Palestinian issue and is actively stimulated both by some governments and by much of the indigenous media. The United States is, thus, held accountable for Israeli actions and is almost universally viewed among these publics as biased on Israel's behalf, including in the United States' international diplomacy (notably, at the UN). How important this factor would be in regard to U.S. efforts to foster the creation of a Persian Gulf security structure is hard to judge. Indeed, even though the Arab Street (along with popular opinion in non-Arab Muslim countries) still regularly becomes inflamed by events in the Israeli-Palestinian conflict, that issue now plays a less-corrosive role than in the past in regard to limiting, politically, the latitude of Gulf Arab states to work effectively with Western states (notably, the United States). There has even emerged a collective Arab position on the so-called peace process that is far from the old "three no's" that emerged during the Khartoum Arab Summit immediately after the 1967 Six Day War.[1] The declaration of the Arab League Summit in March 2002, fostered by Saudi Arabia, still called for Israel to withdraw from *all* the territories it occupied in

[1] There three no's were "no peace with Israel, no recognition of Israel, and no negotiations with Israel" (Palestine Facts, 2009).

June 1967—a condition that Israel will not accept—and included other provisions on refugees and Jerusalem that Israel opposes, but it did, for the first time, offer in return an end to the conflict, a peace agreement, and normal relations with Israel.[2] Indeed, a key component of the Obama administration's approach to Israeli-Palestinian peacemaking is an effort to induce Arab states to build on the 2002 initiative.[3]

Also significant—and derived from the blessing for new thinking provided by the 2002 Arab League Summit—is the fact that there is not the same rigid resistance in virtually all of the Arab world, even beyond those countries that have made peace with Israel, to the possibility that, following a resolution of the Israeli-Palestinian conflict, Israel could become a part of broader efforts to create greater stability in the region as a whole. Notable in this regard was a proposal made by Bahrain's foreign minister, Sheikh Khalid bin Ahmed Al-Khalifa, to the UN General Assembly in September 2008:

> In order for the Middle East to live in a stable and lasting peace, it is incumbent on us to review and re-evaluate our regional outlook, and the possibility of developing new regional frameworks to overcome our longstanding challenges. It is now the time, for example, *to consider the possibility of establishing an organization that would include all states in the Middle East, without exception*, to discuss long-standing issues openly and frankly, in the hope of reaching a stable and durable understanding between all parties. As Arabs, we accept peace as a strategic option, committed

---

[2] Among other provisions, the Arab Peace Initiative of March 27–28, 2002,

> 1. Requests Israel to reconsider its policies and declare that a just peace is its strategic option as well.
>
> 2. Further calls upon Israel to affirm:
>
> I—Full Israeli withdrawal from all the territories occupied since 1967, including the Syrian Golan Heights, to the June 4, 1967 lines as well as the remaining occupied Lebanese territories in the south of Lebanon.
> II—Achievement of a just solution to the Palestinian refugee problem to be agreed upon in accordance with U.N. General Assembly Resolution 194.
> III—The acceptance of the establishment of a sovereign independent Palestinian state on the Palestinian territories occupied since June 4, 1967 in the West Bank and Gaza Strip, with East Jerusalem as its capital.
>
> 3. Consequently, the Arab countries affirm the following:
>
> I—Consider the Arab-Israeli conflict ended, and enter into a peace agreement with Israel, and provide security for all the states of the region.
> II—Establish normal relations with Israel in the context of this comprehensive peace. (Council of Arab States at the Summit Level, 2002)

[3] Seeking Arab willingness to play a positive role in the peace process is, reportedly, a principal motive behind the Obama administration's efforts to get Israel to halt further construction of Jewish settlements in the West Bank. See the following statement from Obama, 2009c:

> And finally, the Arab states must recognize that the Arab Peace Initiative was an important beginning, but not the end of their responsibilities. The Arab-Israeli conflict should no longer be used to distract the people of Arab nations from other problems. Instead, it must be a cause for action to help the Palestinian people develop the institutions that will sustain their state, to recognize Israel's legitimacy, and to choose progress over a self-defeating focus on the past.

to legitimacy, and to concluding past conflicts and hostility, opening instead a new chapter for an historic rapprochement between the peoples of the region towards a better future, dominated by understanding, stability, and prosperity.[4]

Given the clear implication that Israel could be included in a regional framework, this is a remarkable statement. It not only reflects an evolution of politics (albeit on the part of a country that both is relatively inconsequential in Arab politics and has taken less of a hard line toward Israel in the past) but also seems to reflect a rebalancing of objectives, with security within the region as a whole gaining a higher place in the hierarchy of Bahraini interests.

Nevertheless, despite such forward-looking comments, conventional wisdom still holds that the continued lack of resolution of the Israeli-Palestinian conflict inhibits the pursuit of U.S. interests elsewhere in the region, to a greater or lesser degree, including any effort to create viable security arrangements for the region. Conventional wisdom also holds that the European allies expect the United States to take an active role in pushing this conflict *toward* conclusion, if not to a *full* conclusion, as a price of their cooperation in support of U.S. policies elsewhere in the region.[5] Much of this expectation derives from the rising number of Muslims in many European countries; indeed, Islam is the fastest-growing religion in Europe.[6] Also, with the possible exception of Germany, no European country has an affinity for the State of Israel comparable to that of the United States, and the Palestinian cause has, thus, tended to elicit more sympathy from most European governments and publics than from Americans.

In recognition of these two pieces of conventional wisdom and the United States' interests in protecting Israel's security, reducing conflict, and promoting human rights, the United States has continued to be involved in peacemaking, to varying degrees, even after a final peace lost most of its strategic significance for the United States in the prosecution of the Cold War when the Egyptian-Israeli Peace Treaty was signed. These continuing efforts included cosponsoring the Madrid Conference of October 1991;[7] building on Norwegian diplomacy to forge the so-called Oslo Accords of Sep-

---

[4] See Al-Khalifa, 2008 (emphasis added). The minister also reiterated various, more-or-less standard requirements for Israel to fulfill, within the need for a just, comprehensive and durable peace settlement for the Palestinian question, based on ensuring security for all the peoples of the Middle East region, including Israel.

[5] One notable example of this factor at work can be found in the joint press conference of U.S., British, Spanish, and Portuguese leaders in the Azores, four days before the U.S. and Coalition invasion of Iraq. British Prime Minister Tony Blair emphasized the importance of Arab-Israeli peacemaking, most likely because of his domestic political needs. See G. W. Bush et al., 2003. Also, with war on Iraq impending, it was remarkable that President Geroge W. Bush made a statement on Arab-Israeli peacemaking just two days before the press conference in the Azores. See G. W. Bush, 2003. Finally, note that the so-called road map for Arab-Israeli peacemaking was issued by the Quartet parties just at the end of the first phase of the war in Iraq. See Isseroff, 2003.

[6] See, for instance, S. Hunter, 2002.

[7] See the United States and the Soviet Union, 1991.

tember 1993;[8] brokering a peace treaty between Israel and Jordan in October 1994;[9] and convening President Clinton's Camp David II conference in 2000, which led to the so-called Clinton Parameters at the end of that year[10] and thereby produced the best basis for a potential Israeli-Palestinian agreement so far formulated.

The United States has also been engaged in Arab-Israeli peacemaking after 9/11, when, arguably, the strategic requirement for doing so reemerged, this time in order to reduce the capacity of Islamist terrorism's recruiters to cite the Arab-Israeli conflict in their propaganda in the Muslim world. During the administration of President George W. Bush, these U.S. efforts included sponsorship of the so-called Quartet, consisting of the United States, the UN, the EU, and the Russian Federation, which issued its roadmap in April 2003;[11] the so-called Annapolis Process, which was launched at a U.S.-sponsored multinational conference in November 2007;[12] and the relaunching of diplomatic efforts under President Obama.

But how accurate is the conventional wisdom? This is not an idle question; its answer—in each of its parts, regarding the attitudes of both regional actors and the United States' European allies[13]—will help to determine how much effort the United States will need to put into Israeli-Palestinian peacemaking in terms of timing, scope, degree of engagement (and at what political level or levels), and, especially, *extent of success required* to keep the conflict from interfering significantly with other U.S. policies and goals in the region, including any effort to create a new security structure. It is difficult to answer this question except in the process of testing the reverse proposition: a continuation of the Arab-Israeli conflict and either visible U.S. reluctance to be deeply engaged in peacemaking (case 1) or, even with significant U.S. effort, failure to achieve significant results—in reality, not just in presentation—from the standpoint of either regional actors or the Europeans (case 2). If asked, virtually everyone in Arab/Muslim states in the region and in Europe would assert that the United States has to play an active, committed role and (perhaps) also drive the process to closure—although it is true that some Arab governments use the Palestinian issue to divert attention at home away from their own shortcomings and to obscure their unwillingness to share in

[8] See Declaration of Principles on Interim Self-Government Arrangements, 1993.

[9] See Treaty of Peace Between the State of Israel and the Hashemite Kingdom of Jordan, 1994.

[10] See W. Clinton, 2000. These parameters were expanded upon and presented in much greater detail in "The Geneva Accord," a document drawn up by a group of Israelis and Palestinians and presented on December 10, 2003. For the full text, Palestinian Peace Coalition and Geneva Initiative, undated.

[11] U.S. Department of State, Office of the Spokesman, 2003.

[12] See G. W. Bush, 2002c.

[13] Many European governments have become even more anxious that the United States play a vigorous role in trying to bring the Palestinian issue to success because of the significant rise in Muslim immigration to Europe. For this and other reasons, the view that the United States should act decisively in Arab-Israeli peacemaking is prevalent throughout Europe.

responsibility for either the Palestinians' fate or Arab-Israeli peacemaking. But, be that as it many, if anything, European, Arab, and non-Arab Muslim views were reinforced by the Israeli invasion of Gaza in December 2008, the extent of the damage and the level of Palestinian casualties that the invasion entailed, and U.S. tolerance for the invasion.[14]

Of course, many implications flow from such widespread views, including major aspects of the U.S. relationship with Israel, with its foreign policy and domestic political dimensions. But just because both local actors and European allies make these assertions does not mean that they are necessarily controlling. Thus, if the United States demonstrated a continuing strategic commitment to the Persian Gulf region and worked to develop a structure of security that would benefit critical regional and non-regional actors yet failed to achieve what is expected of it in terms of Israeli-Palestinian negotiations, would its security efforts be rejected by those parties simply because of their perceptions of U.S. inadequacy on Arab-Israeli peacemaking?

This is a complex question, but it goes to the heart of considerations regarding the Obama administration's agenda for the Arab-Israeli conflict and the broader Middle East, whether or not the administration tries to foster a new security structure for the Persian Gulf region. Indeed, this would not be the first time that the United States was told it had to do thus and so in regard to the Arab-Israeli conflict if it were to expect cooperation in other matters on the part of either regional states or European allies, only to find that the assertions were more important as declaratory policy than as action policy. In some cases, such political points were made for local consumption and did not prove to be, in American slang, show-stoppers.

Nevertheless, U.S. efforts in the Middle East, overall, are affected negatively by popular attitudes in the Muslim states of the region regarding the U.S. role on the Palestinian issue. Thus, in line with the earlier proposition regarding the Arab Street, U.S. efforts in the Middle East, in general, would certainly *not be made more difficult*—indeed, to some degree, they would definitely be made easier—if the Israeli-Palestinian conflict were settled or at least being pushed in that direction, and the United States is likely to gain standing in the region and with allies if it is actively and seriously engaged in seeking that end. Just how important this is, however, will be impossible to judge except in the event,[15] and the volatility of public opinion in reacting to shocks, whether real (e.g., the 2008 Israeli invasion of Gaza) or perceived (e.g., the false allegations that the United States was somehow involved in the 1979 attack on Mecca) must always be borne in mind. Thus, the better part of wisdom is that, in pursuit of its

---

[14] There was considerable public condemnation of Israel's actions in the Arab and broader Muslim worlds. Some of the Arab governments, however, were less vociferous.

[15] For several decades, some of the Arab states have had an interest in seeing the Palestinian issue continue to fester, within limits. For example, the Palestinian refugee camps could have been eliminated decades ago, with Palestinians resettled elsewhere and pursuing middle-class lives, if the Arab oil producers had been willing to finance this transformation rather than see the camps continue, in part, as a symbol of the conflict with Israel.

security interests throughout the region, the United States needs to be actively engaged in trying to broker what, as a term of art, is called a "just and lasting peace" between Israel and the Palestinians.

As it is, almost from the outset, the new Obama administration took an active role in trying to broker peace between Israel and Palestine,[16] and this included the rapid appointment of a Middle East Envoy, former Senate Majority Leader George Mitchell; meetings between the President Obama and the leaders of Israel, Egypt, and the Palestinian Authority; President Obama's speech to the Muslim world, delivered at Cairo University in June 2009;[17] and President Obama's meeting with the Israeli and Palestinian leaders at the UN General Assembly in fall 2009. The extent to which President Obama obviously sees the importance of pressing the peace process forward can be seen in the fact that, at least initially, he seemed to elevate it over concerns expressed by the government of Israel regarding the Iranian nuclear program and by the fact that he has also pressed Israel regarding Jewish settlements in the West Bank.[18] And, in October 2009, the National Security Advisor, General James Jones, told the first major conference of J Street, in Washington, D.C., "If I could advise the President to solve one problem among the many problems—this would be it. This is the epicenter, where we should focus our efforts. . . ."[19] No doubt, General Jones could not have made a comment like this without the president's approval.

How far the U.S. administration is prepared to press for success in the peace process is not clear, however, especially in view of the current impediments to progress in negotiations and in view of the domestic political costs in the United States of pressing Israel to make significant concessions in the peace process. Again, the issue is the extent to which the United States has to achieve results as opposed to giving the impression that it is doing all it can to achieve them; again, this question cannot be answered in the abstract.

---

[16] A subsidiary issue is whether to put significant effort into trying to broker peace between Israel and Syria.

[17] See the following statement from Obama, 2009c:

> America will align our policies with those who pursue peace, and we will say in public what we say in private to Israelis and Palestinians and Arabs. (Applause.) We cannot impose peace. But privately, many Muslims recognize that Israel will not go away. Likewise, many Israelis recognize the need for a Palestinian state. It is time for us to act on what everyone knows to be true.

[18] For example, see Obama, 2009c: "The United States does not accept the legitimacy of continued Israeli settlements. (Applause.) This construction violates previous agreements and undermines efforts to achieve peace. It is time for these settlements to stop. (Applause.)" It is striking that the Obama administration is also trying to engage a wide variety of Arab countries in the peace process, especially through its efforts to reinvigorate and revise the 2002 Arab League initiative. In addition to the potential value of this approach—although it is still a long shot—to Arab-Israeli peacemaking, it is also a way for the administration to tell Arab states that, if they want U.S. backing for their security, they also have to do something to increase the chance of success of something that the United States is doing, to a significant degree, on their behalf.

[19] See Mozgovaya, 2009.

There is a further aspect of the Arab-Israeli conflict (and, particularly, the Israeli-Palestinian conflict) that also must be factored in: the degree to which different actors in the region attempt to exploit the continuation of that conflict for their own strategic, political, or ideological purposes. It is clearly true that the Iranian government and (especially) President Ahmadinejad are doing so, in addition to any authentic ideological or religious issues limited to the Muslim holy place in Jerusalem that could motivate Iranian attitudes. Especially given that Iran is strategically remote from the zone of the Israeli-Palestinian conflict, Iran's engagement with Hezbollah in Lebanon and, to a lesser degree, with Hamas in Gaza, surely reflects, in major part, its effort to have a political impact in advancing its own interests elsewhere in the Middle East, both vis-à-vis the United States and to affect positively—in theory, at least—its standing in the Arab world. Potential benefits include some level of impact on the Arab Street. In this case, U.S. prosecution of Arab-Israeli peacemaking and, particularly, Israeli-Palestinian peacemaking can help to counter Iran's propaganda efforts and its attempts to put at risk a declared U.S. interest (namely, Israel's sense of security). By the same token, it could be that improvement of U.S.-Iranian relations would reduce Iranian support for Hezbollah and Hamas and thus increase the chance of positive movement in Arab-Israeli peacemaking, including that involving Israel and Palestine.

The bottom line is that the United States would certainly be better off in terms of its overall policies and position in the Middle East if it were deeply and regularly engaged in Israeli-Palestinian peacemaking—a direction that President Obama has begun to take. However, precisely how much and what level of U.S. political engagement cannot be quantified. Certainly, by being so engaged, the United States can help counter the argument made by some in Europe and in the Middle East that they cannot cooperate with the United States on one effort or another because Washington is either not playing its expected role in Arab-Israeli peacemaking or because—in another variant of the argument—Washington is so wedded to Israeli policies that it will fail to be an honest broker.

Thus, in calculating what it has to do in the region, when to do it, and how far to become engaged (and at what level or levels), the U.S. administration will have to judge, at every point, how to balance the price of playing a vigorous and committed role in Arab-Israeli peacemaking, with all of its difficulties and associated costs in U.S. domestic politics, against the price of abstaining (relatively speaking, in both instances). The default option, however, should always be active U.S. diplomatic engagement.[20]

Further, issues relating to Israel's security and to Iran's role in the Middle East, and, especially, the latter's nuclear program—at least as Iran's ambitions and actions are perceived in Israel—would, unless dealt with effectively, vastly complicate any U.S. efforts to foster a new security system for the Persian Gulf. In at least this element,

---

[20] This analysis of the importance of Arab-Israeli peacemaking can be termed *best case*, from the U.S. standpoint. It could well be that the United States will be more strongly pressed to act by Arab or European governments.

therefore, there is a clear linkage between the zone of Arab-Israeli conflict and the Persian Gulf, with diplomacy in each significantly affecting the other. Israel has taken pains to reinforce this point with the United States, regularly pressing for U.S. action regarding the Iranian nuclear program.[21]

But what does even a vigorous effort to promote Arab-Israeli diplomacy mean in practice? In Israeli-Palestinian peacemaking, the devil rests not so much in the details—after all, the Clinton Parameters provide a solid basis for moving forward—but in the nature, character, and development of underlying politics and society. Many Israelis are still not willing to take what for many years have been called "risks for peace." Israeli politics have also not developed to the point at which any prime minister seeking to lead on this issue can expect to receive sufficient domestic political support to prevail. The current Israeli government is not a vigorous supporter of the two-state solution, to say the least, and the Israeli body politic is still adjusting—to use a mild characterization of a profound national unease—after both the 2006 war with Hezbollah in Lebanon, where Israel was fought to a standstill, and the conflict with Hamas in Gaza in 2008. While Israel did prevail in the latter, to the extent that the concept of prevailing can be said to apply, the result did not reassure Israel about the prospects that Palestinians, in general, would be willing to live peacefully next to the Jewish state.

Furthermore, despite all the diplomatic efforts that have been under way for a long time to develop within Palestinian politics, governance, and administration a valid "partner for peace" for Israel, this goal is unlikely to be achieved, whatever is done within the West Bank—where significant political change is also highly problematic[22]—as long as the situation in Gaza continues to fester. It is remarkable that so many observers and even practitioners of Israeli-Palestinian peacemaking act as though Gaza does not exist and can be safely ignored, or at least sufficiently so as to support the belief that it does not have to be dealt with now. This view can arguably be seen as the product of a lack of both vision and political courage to do something about the situation. Since the withdrawal of Israeli forces and settlers from Gaza (which began in August 2005),[23] Hamas' electoral victory in January 2006,[24] and Hamas' coup in June 2007 (when it gained full control of Gaza),[25] the territory has become progressively more isolated.

---

[21] Among many similar statements on the subject, see the following comment made by Israeli Prime Minister Benjamin Netanyahu at the White House in May 2009: "I very much appreciate, Mr. President, your firm commitment to ensure that Iran does not develop nuclear military capability, and also your statement that you're leaving all options on the table" (Obama and Netanyahu, 2009).

[22] However, a number of developments within West Bank politics in 2009 have been pointing in the right direction. See, for instance, Kershner, 2009.

[23] See, for instance, Morley, 2005.

[24] See Wilson, 2006.

[25] See "Hamas Coup in Gaza: Fundamental Shift in Palestinian Politics," 2007.

It is difficult to believe, however, that the Palestinian Authority will be able to form a government able to represent all Palestinians in the occupied territories or to negotiate a settlement with Israel that would meet Israel's minimal requirements (much less those of the Palestinians) if Gaza is left out of the process. It is also most unlikely that there could be any reconciliation or even a workable compromise between the authorities in the West Bank and the authorities in Gaza as long as Hamas holds to its current views or, to put the point more starkly, as long as Hamas is in effective control of Gaza, its residents, its economy, and its politics. This argues for efforts to try weakening the control that Hamas exerts. Even though the number of rocket attacks on Israel from Gaza declined significantly in 2009, the Israeli attacks of December 2008 still did not produce political change in Gaza, which is necessary to the prospects for Israeli-Palestinian peacemaking. This calls into question the utility of the military option for promoting political change. The alternative is for there to be efforts to provide the people of Gaza with a reason to begin shifting political allegiances, and that process would include having a source of economic support in their own lives that does not flow primarily from Hamas.

This seems a simple point, but it was not followed in regard to Gaza, in a sufficiently serious way, either during or after Israeli occupation or, indeed, during the period of Egyptian control of Gaza (1949–1967). The ideal time for a major infusion of material aid to Gaza was August 2005–January 2006, between the withdrawal of Israeli forces and settlements and the legislative election, but it was not provided on anywhere near a sufficient scale. It should have been no surprise that Hamas, the principal source of economic support for Gaza residents, won that election. Unfortunately, that outcome led to a reinforcement of Gaza's isolation, with equally predicable, negative results. This isolation contributes to Gaza's becoming a fertile ground for Islamist terrorism's recruiting agents, with implications for the entire Middle East and, perhaps, beyond.

What is needed now, therefore, is massive, external economic and humanitarian support to Gaza and its people that could begin to weaken the hold of Hamas and begin to reduce the frustrations that help breed terrorism. A usual objection—that aid would be diverted either to the pockets of Hamas' leadership or would be used to strengthen its position—lacks credibility: Aid and investment efforts that were large enough and that were accompanied with a loosening of the Israeli grip on the enclave would almost certainly begin to have the desired effect. Nor should sources of funding be lacking on the part of governments truly interested in working toward a resolution of the Israeli-Palestinian conflict.[26]

If these steps proved fruitful, Israel would be more likely to gain a valid Palestinian negotiating partner that could both work toward peace and deliver on an agree-

[26] For example, in February 2009, some Arab states of the Persian Gulf, led by Saudi Arabia and Qatar, made pledges of help for Gaza, reportedly in the amount of $1.2 billion. See "Gulf States Launch Arab Aid Plan to Rebuild Gaza," 2009. Even if fulfilled, these pledges are only a fraction of what is needed.

ment that would be reached, in part, because that Palestinian partner would represent a collective willingness (on the part of residents of both the West Bank and Gaza) to make peace.

At the same time, from the opposite perspective, leading Arab states are waiting to see both whether Israel will be prepared to take critical steps and whether the U.S. president is prepared to run his own risks for peace, denominated, in part, in terms of U.S. politics. Although the statement might have been made, at least in part, as a matter of bargaining tactics, the Saudi Foreign Minister did say during a July 2009 press appearance with Secretary Clinton that "incrementalism and a step-by-step approach has not and—we believe—will not achieve peace" and that the 2002 Arab Peace Initiative (with its requirement of "full Israeli withdrawal from all the territories occupied since 1967") should be the basis for negotiations.[27] This statement was not a hopeful sign of Arab willingness to meet President Obama half way.

In addition, judging from all past negotiating experience, the U.S. president cannot press for serious peace efforts without prior progress within Israeli and Palestinian politics and society, including an end to Gaza's isolation. Nor will he be able to count on support from Arab states until he shows his own commitment to move forward, not just on tactical issues, such as the Israeli settlements, but also on the big issues that have eventually to be resolved.

There is one further aspect of Arab-Israeli peacemaking that will have a significant impact on the possibility of developing a regional security structure for the Persian Gulf.[28] This is whether, for the United States to consider including Iran in a regional

---

[27] H. Clinton, 2009. Furthermore, Foreign Minister Saud said,

> [t]emporary security, confidence-building measures will also not bring peace. What is required is a comprehensive approach that defines the final outcome at the outset and launches into negotiations over final status issues: borders, Jerusalem, water, refugees and security.
>
> The whole world knows what a settlement should look like: withdrawal from all the occupied territories, including Jerusalem; a just settlement for the refugees; and an equitable settlement of issues such as water and security. The Arab world is in accord with such a settlement through the Arab Peace Initiative adopted at the 2002 Arab Summit in Beirut which not only accepted Israel, but also offered full and complete peace and normal relations in exchange for Israeli withdrawal from all Arab territories occupied in '67. This initiative was adopted unanimously by the Islamic countries at Makkah Summit in 2005. Today, Israel is trying to distract by shifting attention from the core issue—an end to the occupation that began in '67 and the establishment of a Palestinian state to—[sic] incidental issues such as academic conferences and civil aviation matters. This is not the way to peace. Israel must decide if it wants real peace, which is at hand, or if it wants to continue obfuscating and, as a result, lead the region into a maelstrom of instability and violence . . . . (Quoted in H. Clinton, 2009)

[28] A further strand of thinking is that success in Israeli-Syrian peace negotiations would reduce Syria's engagement with Iran, perhaps including a reduction in the passage of arms and other support to Hezbollah in Lebanon. This reduction would make it easier to isolate and pressure Iran. In theory, this could also, in time, cause Iran to be more amenable to some regionwide security arrangements, assuming that its efforts to be the hegemonic power had been scotched. This theory begs a lot of questions, however, including whether Syria would be willing to compromise without an Israeli-Palestinian agreement and whether Israel would be willing both to evacuate settlements on the Golan Heights and to compromise on the issue of the ownership of the eastern bank of the Sea of Galilee (also known as the Kinneret and Lake Tiberias).

security structure (which begs the question whether Iran would be prepared to join under terms that would make sense both to its neighbors and the West), it would first be necessary to deal effectively both with Israel's keen security concerns about a possible Iranian nuclear weapon and with Iran's relationship with Hezbollah and Hamas. Indeed, it is clear that there is linkage, in political fact, between Israeli-Palestinian peacemaking (or lack thereof) and the course of Iranian policies and relations with the outside world and, certainly, the U.S. approach to Iran. In the final analysis, it is doubtful that Iran would be able to block progress between Israel and the Palestinians if the Palestinians were prepared to reach a peace agreement.[29] But Israel's concerns about Iran would still likely be sufficient to keep the United States from being able to test the possibilities of Iranian support for a regional security structure without at least Israel's principal concern, the Iranian nuclear program, having been dealt with adequately from the perspective of the United States and probably also Israel.

This is a clinching argument: To the extent that a new security structure for the Persian Gulf were to seek *to have Iran in* instead of *assuming that Iran would be out* or even *making Iran a country against which the security structure would be directed*, if only as a form of existential deterrence, Arab-Israeli peacemaking must proceed apace.

---

[29] As noted in Chapter Four, in at least one interview that occurred while the September 2008 UN General Assembly was ongoing, President Ahmadinejad ceded primacy on this issue to the Palestinian people:

> Ahmadinejad was asked: "If the Palestinian leaders agree to a two-state solution, could Iran live with an Israeli state?" This was his astonishing reply: "If they [the Palestinians] want to keep the Zionists, they can stay . . . . Whatever the people decide, we will respect it. I mean, it's very much in correspondence with our proposal to allow Palestinian people to decide through free referendums." Since most Palestinians are willing to accept a two-state solution, the Iranian president is, in effect, agreeing to Israel's right to exist and opening the door to a peace deal that Iran will endorse. (Tatchell, 2008)

CHAPTER SEVEN

Regional Tensions, Crises, and Conflicts

Security, as the term has so far been used here, is assumed to be about external threats or challenges to regional countries—e.g., from Iran or from terrorist groups, such as Al Qaeda—and about internal strife and conflict in Iraq. But any security structure worthy of its name also has to take account of tensions, crises, and the possibility of conflicts between individual countries in the region, including members of any new formal security structure that is developed. This work has already introduced the security issue that currently exists in regard to the threats posed by the PKK in Iraqi Kurdistan to Turkey and both Turkey's possible military responses to those threats and its concerns about potential developments within Iraqi Kurdistan (e.g., a declaration of independence). It has also introduced the security issue posed by MEK efforts to destabilize the Islamist regime in Iran. Clearly, for either Turkey or Iran to be willing to participate in a security structure for the Persian Gulf region, their specific concerns on these two scores—and, in Iran's case, others, including internal destabilization fostered from outside—would have to be addressed and dealt with adequately.

But what if other situations develop in which one or another regional country feels threatened by state actions or state-tolerated actions from within the region? For example, relations between Saudi Arabia and Qatar (and, to a lesser degree, Bahrain) have not always been cordial. The same has been true, historically, between Saudi Arabia and Yemen, an issue that has again come to the fore, and there are also occasional stresses in Saudi-UAE relations. The Iraqi government keeps a wary eye out for potential interference from different quarters, and not just from Iran.[1] It is possible, as well, that relations among members of the GCC[2] could deteriorate for one or another

[1] See, for example, "No More Gestures to Saudis: Iraq's Maliki," 2009, which states the following:

> Saudi-Iraqi relations are at a low ebb and Baghdad has no intention of making goodwill gestures because Riyadh sees them as a sign of weakness, [Iraqi] Prime Minister Nuri al-Maliki said on Thursday. Ties have been strained since the U.S.-led invasion of 2003 toppled dictator Saddam Hussein and ended more than 80 years of Sunni Arab domination of Shiite-majority Iraq since the modern state was founded. Maliki's Shiite-led government accuses Riyadh of not doing enough to stop its citizens crossing the border and joining the mainly-Sunni insurgency that has killed thousands of Iraqis in the past six years.

[2] For example, the sizeable Iranian population in Dubai could pose security problems in the future, depending on what else is happening in Iranian relations with Persian Gulf Arab countries and on what threats are posed

reason, including internal political changes that transform the dynamics of interstate relations.[3] In addition, depending on the reach of a new security structure, developments in regard to Syria, Lebanon, Jordan, and even Egypt could have a significant impact in several dimensions.[4] These could include the possibility of preventing even the creation of a viable security structure.[5] In any event, individual candidates for participation in the structure, whether it were organized from within the region or with an active external leader, will, understandably, want to know that the structure will provide reassurance against a range of potential threats and challenges from within the region.

At the same time, some governments considering whether to join a formal security structure could seek support against *internal* political change. It is one thing, of course, to require that neighbors pledge not to engage in subversion or to countenance the activities of subversive elements from their territories. It is quite another thing to try creating a regional security structure in which each member is bound to come to the assistance of another member whose government is being challenged from within.[6] Such assistance is not unknown, of course; among other things, it is the essence of counterinsurgency.[7] But to try writing that requirement into a broader regional security structure, with concomitant political validity and, hence, credibility, could be to put more weight on the arrangements than they can bear. And, even if the United States or other external countries opposed internal change (especially internal change imposed through force) in a regional member country, they would be most unlikely to agree to act as part of a formal compact. The game, in this case, would likely not be worth the candle for just about any member of a formal regional security structure. Trying to institute such an agreement would be antithetical to general provisions

to Iran. Indeed, if there were a U.S. or Israeli attack on Iran, Iranian expatriate communities in the region could become involved in retaliatory actions against Western assets.

[3] Such a development in one or another GCC country—e.g., the overthrow of one of the royal houses—may seem far-fetched, but the possibility cannot be entirely ignored. Saudi Arabia, for example, is not free from internal tensions, as proved by a number of events since approximately 1979, some of which have been directed outward—e.g., against U.S. targets, such as the Khobar Towers—rather than internally.

[4] Major political changes in either Jordan or Egypt—e.g., toward some form of radical Islamist government—would also have a profound impact on Israel and Israeli-Arab relations, which could dwarf other security concerns in the Greater Middle East.

[5] Chapter Six's discussion of the role of the Arab-Israeli conflict and, in particular, Israeli-Palestinian relations introduced one such set of issues that could influence whether a viable security structure for the Persian Gulf region could even be created.

[6] This set of issues was considered at the time of the first post–Cold War expansion of NATO membership. The allies rejected the idea of guaranteeing a democratic form of government in any member state, but this concern added to the care with which aspirants to membership were judged. Thus, one of the informal criteria for membership was demonstrated progress toward democracy, including the holding of more than one democratic election at the national level.

[7] For a comprehensive presentation on counterinsurgency, see Gompert et al., 2008a.

of noninterference in internal affairs, and the agreement could not, in any event, be enforced, although there could be provisions for expelling a country that did not meet an agreed standard (e.g., because it fostered or abetted an insurgency in a neighboring country or worked actively to overthrow its government). Further, one aspect of trying to strengthen security in the Persian Gulf region should be to promote the evolution of societies in the direction of becoming more stable over time, an evolution deriving from a process of economic, political, and social development.[8] This begs the question, however, of whether such transformations will contribute to stability or to instability, at least in the short-to-medium term. It also begs another question about how such transformation is to come about and over what period—an area of effort where, historically, fools tend to rush in, with, at best, mixed results.[9]

---

[8] In the region of the Persian Gulf, this is not a standard that is likely to be incorporated into the formal statutes (if any are developed) of any regional security structure. Indeed, many of the states whose participation would be necessary in order for the security structure to have a chance of succeeding do not have democratic governments. This is a case in which the pursuit of stability, as traditionally defined, would trump internal political characteristics, at least initially, even though the nature of those characteristics could have an impact on the willingness of a member state to respect the security needs of its partners.

[9] A good example of unintended consequences in this area was the elections held in Gaza in January 2006, which were won by Hamas. This result, which many observers did not expect, was clearly unwelcome to Israel, the United States, and some other countries. The United States, which had pressed for the elections, was put in the embarrassing position of opposing the results of elections it had called for. Indeed, there is a much-observed distinction between *democracy of process* and *democracy of results*. The former puts weight on what is being done and how it is being done, regardless of outcome. The latter validates the democratic process only if the right people or party wins. In his initial comment on the election results, President George W. Bush praised the process; only later did he emphasize that the result was unwelcome. Another concern about the outcome of some elections, of course, is contained in this ironic slogan: "one person, one vote, one time."

CHAPTER EIGHT

The Roles of Other External Actors

This discussion has focused almost entirely on the roles that the United States could play in any new regionwide security structure for the Persian Gulf. As the discussion has made clear, however, one reason for analyzing alternatives for such a structure is precisely to determine whether the United States will be able to reduce both its exposure and its responsibilities in a manner consistent with protecting its security and other interests. This assessment excludes the possibility that U.S. leaders might seek to retain, through various instruments, a dominant position in the region for reasons that extend into the realm of what might be characterized by different observers as a form of imperialism, neocolonialism, paternalism, or, simply, a prophylactic approach against some unspecified future danger, but one that could have major consequences if it came into being.[1] That alternative, seeking a dominant position, needs to be explored thoroughly, however: Some observers might argue that the course of U.S. engagement in many parts of the world, including in the Middle East, is a form of "soft imperialism"—an effort to play the role of principal arbiter in regional developments in areas of key interest and concern to the United States. Even in Europe, the United States has always kept a watchful eye on regional developments, although it is, at least on paper, completely committed to the development of the EU, a friendly potential regional hegemon.

Barring the emergence of a U.S. ambition to be the dominant country in the region of the Persian Gulf for its own sake rather than in response to some calculus of security interest, the United States will certainly benefit from sharing whatever role it plays with other external powers. Indeed, since the onset of the serious difficulties that followed the 2003 invasion of Iraq, the United States, under both the George W. Bush and Obama administrations, has already moved significantly in the direction of seeking support from allies and partners in the Middle East and Southwest Asia. Even so, it has still not reached the point of trying to define new security arrangements for the

[1] Even without going so far as to see imperialist or neocolonial motives, there has been support in some quarters for developing modern forms of mandate or trusteeship that follow the models created by the League of Nations and the UN, respectively, for different parts of the Middle East, including the Palestinian territories, Lebanon, and Iraq.

Persian Gulf region that would be based on others' ideas as well as its own and that would entail its ceding a fair degree of leadership, decision, and political influence to others, including European allies.

The Europeans

The idea that the United States might seek to gain the support of European allies and partners in helping to construct, direct, operate, and even guarantee regional security arrangements for the Persian Gulf and environs seems to presuppose that it is the United States that is most interested in what happens there. Of course, that is not true—or, at least, it should not be true. From the moment the initial phase of the Iraq War came to a close in May 2003, the old system of security in the region was shattered, and the United States had no choice, given its own interests, but to take the lead in devising some alternative to put in its place. But the Europeans were also placed on the hook—on this occasion, many of them would argue, through no fault of their own. Indeed, a number would argue that they had opposed the U.S.-led invasion of Iraq precisely because they feared the United States would end up exactly where it did in May 2003: with a potentially reduced capacity to do other things that the Europeans, especially the United States' allies, wanted the United States to be able and willing to do elsewhere in the world that would also help to meet their security and other interests. This is a form of opportunity cost that the European allies experienced because of U.S. engagement in Iraq.

Most or perhaps even all European states might accept that they cannot stand totally aloof from what happens in the region of the Persian Gulf, although some (such as Norway) are not dependent on its hydrocarbons, some do not have significant Muslim populations that are affected by Islamist terrorism, and many do not believe that they suffer from the penalties in the Arab world of an overly close relationship with Israel. Even so, this does not mean that all the European states see eye to eye with the United States on the extent to which a threat or challenge to their interests emanates from the region. Nor is there necessarily agreement on the *nature* of the threat or challenge, even when one is acknowledged, or on the potential *remedies*, even when there is common recognition that every Western state's interests are engaged and that something needs to be done. Nevertheless, it is certainly clear that, as the old saying goes, it is not just the U.S. end of the transatlantic boat that is leaking.[2]

Indeed, over the last few years, there has been slow movement toward agreement on both sides of the Atlantic not just that no one is immune from the bad consequences

[2] Even if the United States were more "in the soup" than its European allies, an American might note that, twice in the 20th century (1917 and 1942), the United States became involved in European wars that it had not started and that Europeans should, thus, be a bit sympathetic to U.S. concerns even if they opposed the original U.S. invasion of Iraq.

of negative developments in the region of the Persian Gulf but also that no one country—i.e., the United States—can reasonably be required to bear a vastly disproportionate share of the responsibility to act. This dawning conclusion can be seen, in some measure, in the European Security Strategy adopted by the EU's European Council in December 2003[3] and in various declarations of the NATO Alliance, such as the Comprehensive Political Guidance adopted at the Alliance's November 2006 summit in Riga, Latvia.[4] Long gone are the old arguments about whether NATO could or should act "out of area."[5] Not only is the Alliance now fully committed and engaged in Afghanistan[6] through every ally's deployments in ISAF:[7] NATO has also formally abandoned (in theory, if not in practice) the idea of placing geographical boundaries on its potential activities. This development is, of course, a long way from widespread understanding in Europe that its individual countries, plus the EU and NATO, have a good deal at stake in a reasonably stable Persian Gulf region—stability that would include the success, in broad terms, of U.S.-led coalition involvement in Iraq.[8] But basic Alliance principles of risk- and burden-sharing, as well as the ambitions of the EU's fledgling foreign and defense policies, argue for European countries to give serious consideration to taking part in the new security arrangements.[9]

---

[3] See European Union, 2003. The strategy posited five threats that are not all that different from those seen by the United States: terrorism, the proliferation of WMDs, regional conflicts, state failure, and organized crime.

[4] Riga Summit Declaration, 2006. Among other things, it states that "the Alliance needs to focus on: a. strengthening its ability to meet the challenges, *from wherever they may come*, to the security of its populations, territory and forces . . ." (emphasis added).

[5] Technically, this means operating beyond the area formally designated in the North Atlantic Treaty, 1949, Article 6. Of course, Article 6 only defines the area in which the treaty's Article 5 applies—the "all for one and one for all" cause. It does not in any way constrain the alliance from choosing to act militarily, or otherwise, elsewhere. This point was debated within the alliance and resolved in the mid-1990s as NATO was considering whether to intervene in Bosnia.

[6] One reason that all the allies deployed security personnel to Afghanistan was the reluctance (shared by most of them) to be involved in Iraq and their desire that the United States not believe that they were shirking all responsibility for regional security, especially in an area that was the locus of planning and training for the 9/11 attacks.

[7] Of course, not all allies are committed in practice, as opposed to in principle, to support NATO-led military actions in Afghanistan. Not only is the most vigorous fighting being undertaken by a handful of allies—notably, the United States, the UK, the Netherlands, Canada, and, increasingly, France—but many of the allies impose so-called caveats on where and when and how their deployed forces can engage in combat. This has become a central issue within the NATO Alliance.

[8] Widespread European schadenfreude over the United States' predicament in Iraq is an understandable but foolish luxury.

[9] Thirteen NATO allies, plus Ukraine, are currently involved in the NATO Training–Iraq (NTM-I), which not only helps prepare Iraqi security forces to take charge of their own future but also shows the United States both that more allies than just those that have sent combat troops are prepared to be helpful and that NATO "has value." NTM-I is an illustration of ways in which outsiders could support an emerging formal security structure in the region. See North Atlantic Treaty Organization, undated-f.

Among other things, this slow growth of a sense of shared responsibility in the Persian Gulf is part of an implicit bargain that has been emerging in recent years: *The United States will continue to be deeply engaged in European security, as a European power, but it needs support from its allies to help meet U.S. interests (and what the United States also believes are European interests) in the Middle East region*, as far east as Afghanistan and even Pakistan.[10] In the case of Afghanistan, the issue is the increasingly problematic commitment of military capacity, in practice, by many of the European allies despite the rhetoric of unity and the pledges made by several allied countries at NATO's December 2009 Force Generation Conference to send up to 7,000 more troops to Afghanistan.[11] Europe's dependence on the United States' continued engagement in European security issues was underscored by the Russo-Georgian conflict of August 2008 and consequent fears in much of Europe about possible Russian actions closer to the European homeland, including against Ukraine and the Baltic states or in the functional areas of energy and cybersecurity.

Furthermore, in many parts of the Persian Gulf region, Western countries other than the United States—and, in some places, also other than the UK and France, which are former colonial powers in the Middle East—can be more effective in dealing both with governments and with populations. This is likely to be so precisely because these

---

[10] It is possible that some of the European allies would be more willing to become engaged in a security structure for the Persian Gulf than to take risks in Afghanistan, to say nothing of Pakistan. Engagement in Afghanistan definitely entails the risk of casualties, while engagement in a security structure would do so only if the structure failed to achieve its basic purposes. European support for such a structure and, as pertinent, involvement in its operation could be a way of "punching a ticket" or "compensating" to meet U.S. expectations in order to increase European claims on the United States to act in ways that are aligned with European interests closer to home (e.g., dealing with the future of Russia). This is a linkage that cannot be easily made, but it is not impossible. The irony here would be that most allies who were reluctant to go to Iraq in 2003 but did go to Afghanistan might now be more willing to go to Iraq (although to perform noncombat roles there).

[11] At the NATO summit in Strasbourg-Kehl in April 2009, the allies agreed to the following:

> Our common security is closely tied to the stability and security of Afghanistan and the region: an area of the world from where extremists planned attacks against civilian populations and democratic governments and continue to plot today. Through our UN-mandated mission, supported by our International Security Assistance Force (ISAF) partners, and working closely with the Afghan government, we remain committed for the long-run to supporting a democratic Afghanistan that does not become, once more, a base for terror attacks or a haven for violent extremism that destabilises the region and threatens the entire International Community. For this reason Afghanistan remains the Alliance's key priority . . . . (Summit Declaration on Afghanistan, 2009)

Nevertheless, the allies collectively agreed to increase their force deployments by only 5,000 troops, 3,000 of which would be there only temporarily—not a ringing endorsement: "[W]e have agreed to . . . assist and support the Afghan National Security Forces . . . [to] secure the upcoming electoral process by temporarily deploying the necessary election support forces" (Summit Declaration on Afghanistan, 2009). See also Evans, 2009. A few days after President Obama's December 1, 2009, speech on the way forward in Afghanistan and Pakistan, the allies agreed, at a NATO force-generation conference, to send about 7,000 more troops to Afganistan. At the London conference on Afghanistan on January 28, 2010, the total number of additional forces pledged by NATO allies was raised to 9,000. This includes an extra 500 German troops plus another 350 troops as a "flexible reserve." See United Kingdom Foreign & Commonwealth Office, 2010; "Afghan Reinforcements: Germany Pledges 500 Extra Troops Plus Big Aid Increase," 2010.

Western countries do not have a colonial past or because, unlike the United States (the Western country that most recently led an unprovoked attack on a regional country—Iraq, as odious as the regime of President Saddam was—and is a strong supporter of Israel), they do not have a past as the legatee of others' colonialism. Furthermore, given some of the activities in which NATO and the EU have already become engaged in different parts of the overall region, for European states to take a greater role, even within a formal security structure, might not prove to be as much of a leap as it would be if arrangements had to be fashioned out of whole cloth.

Like the Americans, the Europeans will have to consider, in addition to playing mentoring, training, nurturing, and diplomatic roles, the precise value of engaging directly in any formal regional security structure versus providing an over-the-horizon presence for any military activities and participating in nonmilitary activities, at which many European countries excel. Various European states are also well placed to engage in training local security personnel and otherwise to be involved with local parties without incurring the same degree of opprobrium often experienced by American government personnel (military, more often than civilian) in particular countries.[12] Certainly, the United States *should* welcome the involvement of as many European governments as possible, with their wide range of capacities that span the full spectrum of power and influence, even at the price of ceding some primacy, sharing political influence in local countries, and accepting a considerable degree of joint decisionmaking. A harbinger of such a development, with implications for European engagement in a possible future regional security structure, was France's opening of a small (up to 500-person) military base—dubbed "Peace Camp"—in Abu Dhabi in May 2009.[13] In fact, it may be that any regional security structure in which the United States plays an important role would only have a chance of succeeding if that role were at least partially masked by the engagement of other Western nations. The term *Western* here includes countries in Asia, as well: For example, Australia, New Zealand, Singapore, and South Korea have at different times been involved with ISAF in Afghanistan, as have Jordan and UAE.[14] Japan, meanwhile, assumed lead responsibility for the Afghan disarmament, demobili-

---

[12] One of the arguments for a drawdown of U.S. military personnel in Iraq is that they could no longer serve as lightning rods for insurgents or be used as an excuse by terrorist recruiters.

[13] President Nicolas Sarkozy said that this base illustrates "'the responsibilities that France, as a world power, intends to assume alongside its special partners in a region that is a nerve center for the entire world'" (quoted in Cody, 2009). France also has a defense agreement with Qatar and with Kuwait (see "Defense Agreement Signed Between Qatar and France," 1998; "France, Kuwait Sign Defense Agreements," 2006). Of course, this French initiative in Abu Dhabi also raises a classic issue in allied engagements in different parts of the world in terms of competition for arms sales and, possibly, other forms of influence, political as well as commercial. See "France Opens Base in Abu Dhabi," 2009.

[14] In addition to NATO allies and partners, the following countries had forces in ISAF as of December 22, 2009: Australia, Jordan, New Zealand, and Singapore (International Security Assistance Force—Afghanistan, 2009). The UAE has also been involved with NATO in Afghanistan. See Rasmussen, 2009c; North Atlantic Treaty Organization, undated-e; "Japan to Dispatch Reconstruction Team to Afghan," 2009.

zation, and reintegration program in 2003.[15] Indeed, there would be significant virtue in engaging non-Western external states in cadres helping to support a regional security structure for the Persian Gulf, provided, of course, that some necessary political limitations were observed.

Other Key External Powers

In considering the future of the Persian Gulf region, it is not possible to ignore the potential engagement and, in some cases, actual engagement of other external powers. Certainly, this is true with regard to the supply of the region's hydrocarbons because the rising economies of East Asia—principally, China and India—are increasing their requirements for the secure supply of energy from the region of the Persian Gulf and from Central Asia (a demand that also requires transit routes). It may not be long before these countries also evince broader interests in the Persian Gulf region, a fact that is already becoming apparent in terms of economics. Whether the flag will follow trade is not a given, of course, but, as both China and India become more active in global affairs, it is reasonable to assume that they will look more than before to the Persian Gulf. Furthermore, the two countries will consider their respective policies regarding both the region as a whole and individual regional countries at least, in part, in terms of other aspects of their relations with the United States as well as in terms of overall great-power ambitions.

Indian competition with Pakistan and incipient competition with Iran are already leading New Delhi to look westward; however, increased Indian involvement in the Persian Gulf would likely be looked on most unfavorably by Pakistan, as Pakistan is already responding to growing Indian involvement in Afghanistan, and this could exacerbate already existing tensions between the two countries.[16] Due to its general

---

[15] See Ministry of Foreign Affairs of Japan, 2009. Also see United Nations Development Programme, 2009. The new Japanese prime minister, Yukio Hatoyama, who assumed office in September 2009, pledged to end Japan's naval refueling mission in the Indian Ocean, a mission that has supported the allied campaign in Afghanistan (see Ito, 2009). But, after a meeting with President Obama, Prime Minister Hatoyama said, "[I] expressed my ideas of our contribution to the issue of Afghanistan and Pakistan, and my ideas on the issue of North Korea. And we will proceed in dealing with these issues in a cooperative manner" (Obama and Hatoyama, 2009).

[16] Pakistan may also become more deeply interested in the Persian Gulf, and its attitude toward a regional security structure there would need to be taken into account. Iran, in particular, could have difficulties with any direct Pakistani role in security issues relating to the Gulf. A large number of Pakistanis have already migrated to Gulf countries, most in search of employment. At the same time, a number of regional countries have expressed concerns about developments in Pakistan, and, in 2008, they joined a new Friends of a Democratic Pakistan Group:

> The Friends of a Democratic Pakistan Group was founded in New York on 26 September [2008]. It consists of the U.S., Britain, Pakistan and the United Arab Emirates as co-chairs, as well as Saudi Arabia, Germany, Italy, France, Japan, Turkey, China, Australia, the EU and the UN. The group's objective is to mobilize greater international support for Pakistan in the light of the country's precarious situation. (Auswärtiges Amt, 2008)

desire to be taken more seriously on the world stage, India has tested its capacities to project naval power into the Arabian Sea and the Persian Gulf.[17]

This interest on the part of China and India does not mean that either needs to be included in a regional security structure, at least not in its early stages. However, the United States and others with concerns about the Persian Gulf and the future of China and India do need to take into account those two states' interests, the evolution of their overall policies, and their increasing projections of influence.[18] It is not clear how much China or India would be able to contribute to a regional security structure. By the same token, it is not clear that either would have any incentive, let alone the ability, to confound arrangements agreed by regional states and backed by other outsiders, including the United States, European countries, and such institutions as the EU. Indeed, China and India stand to gain added assurances about the security of hydrocarbon supply from the region, a supply that would be looked after by others in their own interests. Judgments about the potential roles of these two countries in a regional security structure—or simply *in relation to it*—will have to be made as the process of considering and developing a structure moves forward. China and India might see opportunities to be engaged, or, if they are left out, they might see limits on their regional influence or aspirations. Furthermore, attitudes on the part of both India and Pakistan in regard to the Persian Gulf and any security developments there will be strongly influenced by the course of the conflicts in Afghanistan and Pakistan and by the evolution of Iran's relations with the West. Thus, it should be expected that India in particular will want to play an increasingly prominent role in and about the Persian Gulf and, thus, will make choices regarding its stance toward any regional security structure and its sponsors.[19]

This provisional quality of involvement in the region by China and India—even with the latter's preoccupation with Pakistan on its northwestern frontier and with its increasing interest in the overall region that abuts the Persian Gulf—certainly does not apply to Russia, beginning with its proximity to the region and some historical engagements. There is, of course, the whole congeries of issues relating to the supply of hydrocarbons from Central Asia and the Transcaucasus, and a bidding war is already underway in regard to the transit of these commodities: The issue is which pipeline or pipelines will go through which countries. Nor can implications of the Russo-Georgian conflict of 2008 be completely isolated from what is happening in the Persian Gulf region. This is true not just because of geographic propinquity but also because of Russia's newly assertive role on the global stage, or at least on the Eurasian

[17] For example, India sent a small flotilla of six ships to the region in 2004, calling at ports in Oman, Iran, UAE, and Bahrain ("Indian Warships in Persian Gulf," 2004).

[18] For a discussion of Chinese interests and attitudes, see Garver, Leverett, and Leverett, 2009.

[19] Indian calculations about wanting to play a more active role in the Persian Gulf and environs could be affected, over time, by the end-user monitoring agreement that India signed with the United States in July 2009, which will permit high-technology U.S. arms sales to India. See Kessler, 2009.

stage. It is certainly seeking, in general, to return to the ranks of great powers. The 2008 crisis over Georgia was only partly about the two breakaway enclaves of South Ossetia and Abkhazia, the South Caucasus overall, and Russia's evident desire to discipline "upstart" Georgia. It was also the result of Russia's decision to choose that opportunity to make some broader points about its own future role, and, in this respect, it could well have miscalculated.[20]

Russia already plays some roles in the Middle East. It is part of the Quartet for Arab-Israeli peacemaking, although its role is quite limited. Immediately after 9/11, it agreed to grant the United States basing rights in some Central Asia countries from which to pursue operations against the Taliban and Al Qaeda. It later rethought this position, however. Notably, after the Georgia crisis, it clearly played a role in Kyrgyzstan's February 2009 revocation of U.S. air basing rights at Manas, which had facilitated the United States' efforts to supply its forces in Afghanistan.[21] Yet, during President Obama's visit to Moscow in July 2009, Russia agreed to U.S. overflights of Russian territory to resupply U.S. and NATO forces in Afghanistan, including with lethal equipment; in addition to wanting to improve bilateral relations, the Russians may have had in mind an attempt to increase their potential role in the broader region.[22] They also have genuine concerns about the future of Central Asia as well as Afghanistan, especially about the possibility that Islamist radicalism can spread to Afghanistan's neighbors to an extent even beyond that seen in recent years. And they are concerned about the trade in narcotics originating in Afghanistan.[23] Perhaps most remarkable of all have been expressions of Russian concerns that ISAF and NATO will not succeed in Afghanistan. In the words of two prominent Russians, Moscow's ambassador to the NATO-Russia Council and a former Soviet commander in Afghanistan, writing in the *New York Times* in January 2010, "Officials in Brussels and Wash-

---

[20] As the author has argued, Russia stood to lose the most from the crisis with Georgia, given its need for access to the global economy (a need far beyond what either Chairman Vladimir Lenin in the 1920s or Premier Joseph Stalin in the late 1940s ever faced with the Soviet Union). This factor needs to be taken into consideration in judging what role Russia is likely to seek to play in the region of the Persian Gulf and the extent to which it should be invited to be engaged in any developing security structure. See R. Hunter, 2008a; Gwertzman, 2008, quoting the author.

[21] See Bumiller and Barry, 2009. The argument has been advanced that Moscow took this step to divert U.S. dependence away from Kyrgyzstan and toward Russia. Indeed, in February 2009, Moscow agreed that nonlethal cargo could transit Russia to NATO forces in Afghanistan. See Moss and Bokhari, 2009.

[22] The agreement provides for 4,500 overflights per year. See "Russia Approves U.S. Military Overflights to Afghanistan," 2009.

[23] See, for instance, India-Russia Joint Declaration, Moscow, Russia, December 7, 2009:

> They agree that the fight against terrorism cannot be selective, and drawing false distinctions between 'good' and 'bad' Taliban, would be counter-productive. . . . The Sides reaffirm their long-term commitment to a democratic, pluralistic and stable Afghanistan. They are in favour of enhancing the role of the International Security Assistance Forces in combating the illegal narcotics infrastructure in Afghanistan.

ington who are thinking of a rapid exit strategy for the ISAF mission are engaged in elaborating on a suicide plan."[24]

Most consequently for U.S. policy toward the region, at least in the short term, Russia has been part of ongoing negotiations regarding the future of Iran's nuclear programs, and it has exhibited on-again, off-again support for U.S. efforts to impose increasingly strict economic sanctions on Tehran. No doubt, it has its own interests regarding Iran and the region as a whole. Moscow is making broader calculations in regard to Iran that concern Russia's position not just regionally but also toward Europe and the United States. These calculations have led Russia, from time to time, to sell weapons and other military equipment to Iran,[25] with the most-prominent controversy relating to reports of the sale of S-300 anti-aircraft weapons.[26] The Russians no doubt balance their interests in building relations with Washington against currying favor in Tehran.

The issue in this work, beyond considering the future of Russia's diplomacy and other activities, including material supply to Iran and Russian policies on the Iranian nuclear program, is whether Moscow should be included in any new security structure for the Persian Gulf. The question can also be posed the opposite way: *To what extent could such a structure succeed if Russia either stood aloof or were an active opponent?* Certainly, a Russia that is simply passive but supportive, as it has tended to be in the Quartet, is much to be welcomed. Indeed, Russia has indicated a desire to promote regional security, as reiterated by Foreign Minister Sergey Lavrov in August 2008:

> Acting on our belief that there can be no force-based solutions to the existing problems in the sub-region, we lay emphasis on forging collective efforts towards their political settlement with an eye on establishing a security system involving all regional and other concerned parties without exception.[27]

---

[24] They continued:

> Withdrawal without victory might cause a political collapse of Western security structures. . . . [I]f the [NATO] alliance does not accomplish its task, the mutual commitments of its 28 member-states would be undermined and the alliance would lose its moral foundation and raison d'être. . . . A pullout [from Afghanistan] would give a tremendous boost to Islamic militants, destabilize the Central Asian republics and set off flows of refugees, including many thousands to Europe and Russia. (Gromov and Rogozin, 2010)

An alternative explanation for this statement is a possible Russian desire to see the West continue to be bogged down in Afghanistan, or at least not to succeed where the Soviet Union had failed.

[25] See, for instance, Beehner, 2006.

[26] See, for example, Keinon and Associated Press, 2009.

[27] "Interview of Russian Minister of Foreign Affairs Sergey Lavrov," 2008. Minister Lavrov went on to say the following:

> The Russian concept envisages gradual advancement towards this goal, beginning with the solution of the most acute problems, such as the Iraq crisis and Iran's nuclear program. At the same time it is necessary to lead matters in an ongoing way toward all-round improvement of the situation in the sub-region via the elaboration of confidence-building measures, security guarantees and achievement of agreements on a fight against transna-

As for the broader question of formal Russian involvement in any regional security structure, much will depend on whether Iran is inclined to take a positive or negative stance toward that structure and whether, if it were being at least conditionally positive, it would seek to enlist Russia as a friend in court. If Tehran proves inclined to be obdurate, for whatever reason, this could work to the advantage of Moscow, or Moscow could seek to gain ground in its relations with the West by playing the role of interlocutor with Tehran. But if Iran proves inclined to be positive, agreeing to operate within a framework that makes sense to other regional countries and key Western outsiders, then the value to the West of a Russian role would likely be significantly diminished, other than in the context of an effort to improve U.S.-Russian relations overall. The same would be true if the United States and Iran were able to create a viable modus vivendi with one another, possibly leading to limited areas of cooperation (e.g., over Afghanistan). The bottom line, however, is that any efforts to foster the creation of a new security structure for the region should at least begin with exploring possibilities with Russia and seeking to create incentives for its support. With the possible short-term exception of a rise in oil prices because of regional turmoil, Russia does, after all, stand to gain, as do other countries, from a region that is more likely to be stable than regularly in turmoil.[28]

---

tional terrorism and other challenges. The convocation, with favorable conditions for this, of an international conference on security in the Persian Gulf zone with the broadest range of participants would facilitate movement forward along these routes.

[28] This work canvasses several different external institutions that might be considered either for a direct security role in or in relation to a Persian Gulf security structure or as models to be drawn upon. Two such institutions that are not canvassed in this work are the Shanghai Cooperation Organisation (SCO) and the Collective Security Treaty Organisation (CSTO) of the Commonwealth of Independent States. Several states with interests in the Persian Gulf are members of SCO, including Russia and China. Iran, Pakistan, and India are observers. Russia, Armenia, Belarus, Kazakhstan, Kyrgyzstan, Taikistan, and Uzbekistan are members of CSTO. Nevertheless, the difficulties that Western states would have with any direct engagement of either organization in the Persian Gulf region would likely be such to limit their potential as models or participants in a regional security structure that would also need to have Western support. For more information on the two organizations, see the Shanghai Cooperation Organisation, undated; GlobalSecurity.org, 2009c.

CHAPTER NINE

Building Blocks for a Regional Security Structure

In considering the creation of a regional security structure for the Persian Gulf, an important point to ponder is whether there is value in creating formal political and security commitments among various countries in the region and perhaps even in requiring that this be done before other steps are taken. These commitments could take many forms. One common form is *collective security*, which is an "all-against-one" approach designed to provide incentives for all members to support what is agreed by all to be a common good in the interest of common security against any threat from any member (in the case of the Persian Gulf, this threat would be aggression or another actionable threat from any Persian Gulf state that was itself a member of the collective security pact). Such an agreement can impose high requirements in terms of discipline and willingness to act, as witnessed by the developments that led to the collapse of the most famous of all collective security pacts, the League of Nations.

Another popular form of security commitment is *collective defense*, a key example of which is the NATO Alliance. In this form, all the parties agree to come to the aid of any member or members that are subject to some untoward behavior (aggression is the most common trigger) that emanates from outside the pact. This is an "all-for-one-and-one-for-all" approach. It is less demanding in terms of discipline and collective willingness to act because a subset of members can agree to take action (e.g., military action) in the name or in behalf of all even if all do not endorse the action.[1] (The military or other response can also take the form of a "coalition of the willing," which can exist with or without formal treaty commitments.) This form of mutual security commitment tends to imply that, from the outset, there is an "other": one or more nonmembers that are considered to be enemies or at least sufficiently hostile as to be beyond the pale. This approach tends to lead to rigidity in assessing the source of threats to security, if only to provide a politically compelling raison d'être for the security struc-

[1] The NATO Alliance has an even higher standard for making decisions. It operates on the so-called consensus principle, which is that any ally can veto any action. (NATO actually never votes; a "veto" is signified by a country's not "joining a consensus.") Yet, even when NATO decides to take military action (e.g., under Article 5 of the Treaty of Washington), not all members have to take part—Article 5 only requires that each ally take "such action as it deems necessary"—and some may lack the military capabilities needed to participate. Those that do take part then become a "coalition of the willing and able."

ture that is more than a desire to prevent or contain tensions or conflict of whatever nature and from whatever source. It may also prove difficult for such an arrangement later to welcome into its membership the party or parties originally characterized as the "other" without some fundamental change in the latter's behavior. (In regard to NATO membership, this development became possible only after the collapse of the Soviet Union and the dissolution of the Warsaw Pact.)

For purposes of analysis, this work renders no a priori judgment about whether formal treaty commitments would be necessary, at least at first, for the creation of an effective regional security structure. Rather, it lays out a building-block approach, presenting and analyzing some possible alternative approaches to security—setting parameters, as it were—that could, one or severally, be drawn upon in formulating a viable security structure. It may be, in fact, that it would be best to proceed to formal security commitments (for example, in the form of a treaty) only after selections have been made from among these building blocks—or others—and at least the initial work of building the security structure is well under way. Indeed, at that point, a formal treaty commitment could be judged unnecessary to the structure's effective functioning.

This approach of holding off on developing political and security commitments can have particular value in circumstances in which either there is merit in preserving flexibility in terms of future membership (e.g., on the part of an Iran that is not part of the initial efforts) or the difficulties of gaining agreement among the potential parties to the agreement would render mutual commitments, created with sufficient specificity to be effective, hard to achieve. In the Persian Gulf overall, and even just among the Arab states, this last point would seem to apply, at least in today's circumstances. Even if circumstances change (e.g., if Iran acquires nuclear weapons and the Arab states of the Gulf mutually agree that threats to their security have increased significantly), a judgment could be made at that later time about how to proceed; in any event, the potential members' coalescence around formal security commitments would be unlikely to have much security-producing effect without the active engagement of one or more outside military powers, especially the United States.[2]

---

[2] If the security situation in the Persian Gulf were to deteriorate to the point at which the United States would want to provide assurances to various states (e.g., following Iran's acquisition of nuclear weapons), the United States would not likely see much (if any) value in the local Arab states' themselves first banding together in a mutual security pact. This would be the opposite of the situation in Europe in 1949, when the willingness of five Western European states—Britain, France, and the three Benelux states—to create a formal security structure, Western Union, helped to convince the U.S. Congress that they were serious about their own security and, thus, could make a reasonable contribution to what became the North Atlantic Alliance. In the Persian Gulf, by contrast, effective military responses to any serious military threat would almost certainly have to rely on the direct engagement of the United States or one or more of the major European military powers.

Potential Models or Partners

In assessing possible building blocks for a regional security structure for the Persian Gulf, it is worth canvassing alternatives based on experience elsewhere. These potential partners or models are discussed in this section.

As has been true of all security structures created in the modern era, one that can be effective for the region of the Persian Gulf must reflect the particular needs of its members; their histories, cultures, societies, politics, and perceptions, both of the region and of their neighbors; and the broader international milieu within which they live. Thus, each such security structure, including one for the Persian Gulf, must be, almost by definition, unique. There can be no ideal model, no one-size-fits-all approach. Elements of other security structures—and organizations or alliances—can prove instructive in developing the parameters for a Persian Gulf security structure and in crafting its particulars, but these must be analyzed and drawn upon with care in a way that respects inherent differences and does not seek to apply lessons learned from elsewhere too ambitiously—in other words, the analogue of "fighting the last war" must be avoided. At the same time, a security structure that is developed for the Persian Gulf could seek partners from outside the structure or the region itself, whether these partners are institutions or individual countries or groups of countries. Further, there is no particular reason, at least from the outset, that the regional security structure could not draw on the experience of more than one outside institution or set of security relationships. Additionally, there is no requirement that it have only one set of partners or not accept partners that are, in other particulars, at odds with one another but might, for the purpose of promoting interests in common in Persian Gulf security, want to work together or at least each work with the regional structure in parallel.

The following sections present a number of possible models or partners to be considered for a Persian Gulf security structure, each of which needs to be assessed both on its own terms and in terms of the elements within it that might provide useful or instructive for the Persian Gulf.

The North Atlantic Treaty Organization

If the United States were to look for a ready-made alliance structure that could have a major role to play in the Persian Gulf region (in support of local arrangements), it would identify NATO, although doing so would beg a series of questions about whether a role for NATO would be appropriate to regional circumstances and could be accepted by regional countries, in whole or part. Already, NATO has assumed (since 2003) principal lead responsibility, under UN mandate, for ISAF, and all 28 allies have sent military personnel.[3] Also, even though many of the NATO allies opposed the U.S.-led

[3] Iceland, which has no military forces, sends other security personnel. As of this writing, 14 other countries also have troops in ISAF under NATO command. A number of these NATO partners could be part of any NATO involvement in a Persian Gulf security framework.

invasion of Iraq in 2003 and have not wanted to support Coalition military operations in Iraq since then, 12 members of NATO (plus Ukraine, a member of the Partnership for Peace [PFP]) take part in NTM-I, which plays a limited but useful role in training Iraqi security personnel.[4] From time to time, there have also been expressions of interest from some NATO quarters about the possibility of a NATO force's becoming engaged in Palestine after the conclusion of a viable Israeli-Palestinian peace agreement.[5] (This idea is explored in some detail in earlier RAND research.[6]) Even though NATO has taken no formal position on this possibility, the idea has not encountered any active opposition. Indeed, there is a general view that, if it were possible to bring the Israeli-Palestinian conflict to a close, becoming involved in some fashion in helping to make a peace effective, even with deployed military forces, would have significant merit. As of now, of course, no NATO ally (including the United States) has had to face a decision on this question, and, thus, what actual responses would be cannot be accurately predicted.[7]

Most directly pertinent to Persian Gulf security is the fact that, in June 2004, the NATO summit held in Istanbul adopted its Istanbul Cooperation Initiative (ICI). Although the initiative could be opened to all interested countries in the Persian Gulf region that subscribe to its aims and content, in practice, it has so far extended only to four of the six countries of the GCC (Oman and Saudi Arabia have, so far, not elected to join the ICI).[8] Essentially, the initiative's aims will be achieved through practical cooperation and assistance in priority areas, including defense reform and planning, military-to-military cooperation, fighting terrorism, countering the spread of WMDs, promoting border security, and cooperating in civil-emergency planning. Notably, the ICI proposal neither posits a security threat to any of the potential entrants nor names any specific country (including Iran) as a possible threat to regional security or to any

---

[4] Britain, Bulgaria, Denmark, Estonia, Hungary, Italy, Lithuania, the Netherlands, Poland, Romania, Turkey, Ukraine, and the United States. See North Atlantic Treaty Organization, undated-f.

[5] See, for instance, the following comments from a 2005 speech in Israel by NATO Secretary-General Jaap de Hoop Scheffer:,

> [T]he responsibility for achieving peace and stability in the region lies first and foremost with the parties themselves. In that context and within these parameters, the idea of a NATO assistance has been brought up. I have stated many times the necessary preconditions before envisaging any NATO contribution. There would first have to be a lasting peace agreement between Israelis and Palestinians. Moreover, the parties concerned must be in favour of a NATO role in its implementation; and there would have to be a UN mandate. These conditions do not yet exist. (Scheffer, 2005)

[6] R. Hunter and Jones, 2006.

[7] R. Hunter and Jones, 2006. The issue of security in an independent Palestinian state has taken on more currency in light of Israeli Prime Minister Benjamin Netanyahu's proposal for a Palestinian state that would be demilitarized: "[The] second principle [for a peace agreement] is demilitarization. Any area in Palestinian hands has to be demilitarized with solid security measures" (Netanyahu, 2009).

[8] Istanbul Cooperation Initiative, 2004. Kuwait joined in December 2004; Qatar and Bahrain joined in February 2005; UAE joined in June 2005.

country. In theory, ICI could even be open to Iran, if Iran were willing to meet NATO's stated requirement of subscribing "to the aim and content of this initiative, including the fight against terrorism and the proliferation of weapons of mass destruction."[9]

There has not, however, been great enthusiasm on either side of the ICI arrangement. The fact that Saudi Arabia, the most prominent of the GCC states, is still sitting on the sidelines, along with Oman, the state in the region generally most reluctant to compromise the independence of its foreign policy, shows that ICI does not have universal appeal among the most-obvious potential members. Indeed, in much of the Arab Middle East, NATO is viewed as shorthand for the United States, and, where that shorthand involves a military association, involvement with NATO is, thus, resisted in some quarters. Not even all the members of NATO's Mediterranean Dialogue (notably, Egypt) are without doubts concerning the relationship.[10] At the same time, however, the imprimatur "NATO" does and would help some Arab countries in terms of engaging with a major element of Western military security because engaging with NATO is not exactly the same as being involved with the United States, even though the United States is the leading member of the Alliance. There is a lesson here for the parameters of a security structure adapted or created for application directly to the Persian Gulf and environs: The idea of a structure that is widely perceived, by populations even more than by governments, as being "made in the West"—and, with even stronger force, "made in the USA"—could pose some political problems from the start.

For their part, some NATO allies also see ICI as a mixed blessing, a potential forerunner to engagements that could take them—and the Alliance—into deeper waters than they are yet prepared to test. For some, in any event, national motivations for involvement in the region are more related to commerce than security. And many are reluctant to go too far in the direction of getting their feet wet until there is greater clarity on several points: most notably, the future of U.S. engagement in the region after Iraq; the course of Western diplomacy with Iran (and the avoidance of war with it); the course of the Arab-Israeli peace process (and, especially, Israeli-Palestinian diplomacy and the U.S. role in it); and developments in Afghanistan (and now Pakistan), especially the demands placed on several of the allies by the United States to do more militarily then they have been doing. In short, there is clear understanding in Europe about the importance of the Persian Gulf based on a variety of reasons, including oil, terrorism, WMDs, commercial access, and migration to Europe (although

---

[9] Istanbul Cooperation Initiative, 2004.

[10] The members of the Mediterranean Dialogue and their dates of accession are Algeria (March 2000), Egypt (February 1995), Israel (February 1995), Jordan (November 1995), Mauritania (February 1995), Morocco (February 1995), and Tunisia (February 1995). See North Atlantic Treaty Organization, 2004:

> The enhanced Mediterranean Dialogue [adopted at the June 2004 Istanbul NATO summit] will contribute to regional security and stability, by promoting greater practical cooperation, enhancing the Dialogue's political dimension, assisting in defence reform, cooperation in the field of border security, achieving interoperability and contributing to the fight against terrorism, while complementing other international efforts.

most Muslim migrants to Europe come from the Maghreb rather than the Mashreq and the Persian Gulf).[11] But, with a few exceptions, there is not in Europe the same sense of strategic requirement or urgency found in the United States that could lead readily to European allies' engagement in a formal security structure, whether it is based on one or another aspect of NATO (e.g., as an extension of the ICI, in either membership or ambition) or something created *de novo*.

Nevertheless, the new NATO secretary-general, former Danish Prime Minister Anders Fogh Rasmussen, has, in a remarkable development, listed ICI, along with NATO's Mediterranean Dialogue, as one of his three top priorities, second only to Afghanistan and NATO-Russian relations. In his first press conference, he said,

> [a]nother partnership will also be a priority for me: NATO's relationship with the Mediterranean Dialogue [MD] and Istanbul Cooperation Initiative countries.
>
> Let me assure the Government and the people in the 11 MD and ICI countries that I am fully committed to building stronger relations with them, on the basis of mutual respect, understanding and trust, and to face common challenges: terrorism, proliferation, the dangers of failed states. Starting today, I will take concrete steps to engage with the MD and ICI countries[.] I will personally engage in dialogue with all of them, to hear their views, and to help support their reforms. . . . This Alliance has, over years, built up a strong relationship and cooperation with our MD and ICI partners. I will build on this strong foundation throughout my term as NATO [secretary-general] . . . .[12]

Of course, this formulation is a long way from a commitment—or even a hint—that NATO will endorse a regional security structure with significant allied participation. This does not mean that it cannot happen, however, either as an extension of ICI or a new formulation. And, certainly, it does not mean that European countries with traditional ties to the region—especially Britain and France—will rule out involvement in some formal structures even if they cannot convince other NATO nations to take part. Britain has retained significant ties and engagements in the region, especially in arms sales, even since its withdrawal from East of Suez at the end of the 1960s. France is becoming increasingly interested in the region, in part, it appears, as an element of a larger effort to play a wider role on the world stage and perhaps also to engage in some limited competition with the United States while, at the same time, formally returning to full participation in NATO's integrated military command structure and

---

[11] As noted earlier, Muslim migration to Europe is a major cause of European desires to see the United States vigorously prosecute Arab-Israeli peacemaking.

[12] Rasmussen, 2009a.

seeking improved relations across the Atlantic.[13] Notably, in January 2008, President Sarkozy agreed with Sheik Khalifa bin Zayed al-Nahyan, president of UAE, to establish a modest French military base in Abu Dhabi (as mentioned in Chapter Eight).[14] President Sarkozy has been active in trying to broker elements of the Arab-Israeli conflict (his efforts have included engagement with Syria), and he has been promoting a so-called Union of the Mediterranean.[15] He has also been seeking a role for Syria in trying to defuse the issue of the Iranian nuclear program.[16] At least in part, his efforts have reflected a desire to increase the role of the EU—France held the EU presidency during the second half of 2008—and France's leadership role within the EU.[17]

Since October 2003, three members of the EU—Britain, France, and Germany—have been engaged in a series of negotiations with Iran over its nuclear program. These talks have included Solana, in his role as EU High Representative for the CFSP. From the standpoint of issues being explored in this work, these talks have been significant because the three European states, acting in part as proxies for the EU as a whole, clearly understand not just the strategic importance of Iran's potential acquisition of nuclear weapons but also the impact of Iran's role on the broader region, the need to try reducing tensions between the United States and Iran, and the importance of trying to reduce the risk that the United States (or Israel) would attack Iran, at least in circumstances short of some major act by Iran that would make such a response both necessary and proper in terms of preserving the security interests of NATO Alliance countries as a whole. [18]

After the Iraq War, which produced the most-serious divisions within NATO in its history, the view is widespread in Europe that a U.S. attack on Iran could have an impact on the Alliance far more serious than that created by the invasion of Iraq. This view could be exaggerated, but concern about preventing war with Iran is certainly

---

[13] President Sarkozy formally embraced France's return to NATO's integrated military command structure in a speech on March 11, 2009, just weeks prior to the April 3–4, 2009, NATO summit. See Sarkozy, 2009.

[14] Moore, 2008.

[15] See Erlanger and Bennhold, 2008.

[16] See, for instance, Ibrahim, 2008.

[17] The assumption by a senior French officer of the position of Strategic Allied Commander Transformation in Norfolk, Virginia, in September 2009 also indicates a heightened French desire to play a major role within NATO. Allied Command Transformation could become a significant support element for a regional security structure created for the Persian Gulf region that either included some direct role for NATO or the use of some NATO training and planning capacities, even if at arms length.

[18] Notable were the following comments of President Sarkozy in his address to French ambassadors in August 2007:

> The parameters are known; I will not go through them except to reiterate that an Iran with nuclear weapons is unacceptable to me and to stress France's full determination in the current process, which combines increasing sanctions but also openness if Iran chooses to honour its obligations. This approach is the only one that can keep us from facing a disastrous alternative: an Iranian bomb or the bombing of Iran. (Sarkozy, 2007)

prevalent in Europe. The European allies have, thus, been gratified both by the shift in U.S. policy toward Iran under the Obama administration, including its announcement in April 2009 of its willingness to take part in talks between the so-called EU Three and Iran,[19] and by the start of multiparty talks at the beginning of October 2009.

This allied concern about the future of U.S. policy regarding Iran, especially the question of possible military confrontation, could have a side benefit in terms of the potential readiness of European states, whether in NATO, the EU, or—for a few of them—bilaterally, to consider engaging in a venture to develop a security structure for the Persian Gulf region that could, over time, help to reduce the risk of conflict and to increase predictability about developments in the region.

The point must not be pushed too far, however. Even with U.S. commitment and leadership—a sine qua non for any security structure for the Persian Gulf and environs that would deal with serious issues and include the possible use of military force from outside—there is likely to be considerable reluctance on the part of any European state (Britain and, possibly, France could be exceptions) to become engaged, especially because of the long-term commitments that could be involved. However, if the United States did succeed in crafting at least the framework for a security structure for the region, the chance of NATO playing a supportive role—as it is doing, in a fledgling way, with the NTM-I in Iraq—would increase.[20]

Here, as in other aspects of transatlantic cooperation in the times ahead that can loosely be called *post-Iraq*, in contrast to the negative attitudes that emerged in NATO after the 2003 invasion of Iraq, the United States would have to factor into any planning the need to show proper deference to the ideas and attitudes of potential European (or other) partners in a new Persian Gulf venture, for example, through a NATO support role assisting a structure rooted in efforts and organizations that were

---

[19] "If Iran accepts [meeting the representatives of the five permanent members of the U.N. Security Council plus Germany], we hope this will be an occasion to seriously engage Iran on how to break the logjam of recent years and work in a cooperative manner to resolve the outstanding international concerns about its nuclear program" (State Department Spokesman Robert Wood, quoted in "U.S. to Join Nuclear Talks with Iran, State Department Says," 2009).

[20] See this statement from the declaration released in Bucharest, Romania:

> 17. We reiterate the Alliance's commitment to support the Government and people of Iraq and to assist with the development of Iraqi Security Forces. We have responded positively to a request by Prime Minister Al-Maliki to extend the NATO Training Mission–Iraq (NTM-I) through 2009. We are also favourably considering the Government of Iraq's request to enhance the NTM-I mission in areas such as Navy and Air Force leadership training, police training, border security, the fight against terrorism, defence reform, defence institution building, and Small Arms and Light Weapons accountability. NTM-I continues to make an important contribution to international efforts to train and equip Iraqi Security Forces and, to date, has trained over 10,000 members of these forces. Complementing these efforts, NATO has also approved proposals for a structured cooperation framework to develop NATO's long-term relationship with Iraq and continue to develop Iraq's capabilities to address common challenges and threats. (Bucharest Summit Declaration, 2008)

NATO also agreed at its April 2009 summit to create a NATO Training Mission for Afghanistan. See North Atlantic Treaty Organization, 2009b.

homegrown. In shorthand, even with effective U.S. leadership, if allies and partners are to be asked to share risks and responsibilities, they will also require a fair share of decision and influence beyond even the formal requirement for unanimity for all decisions taken by the North Atlantic Council. *It is hard to overstate this point, especially since most of the allies still see their security interests as mainly rooted in Europe even when they subscribe, formally, to broader formulations about NATO's reach beyond the European continent.*[21] Afghanistan has been a most-chastening experience.

But this requirement for sharing decisionmaking and influence with allies should not, in fact, be seen by the United States as a burden or serious limitation—indeed, quite the opposite. One essential quality of any new security structure for the Persian Gulf and environs is that it has broad appeal and support, not just among regional countries (the hard part) but also among supporters from outside (the relatively easy part). A shared sense that a security structure, however imperfect, is better than the alternative—e.g., the likelihood of conflict or a generally deteriorating security situation, over time—should help to bolster the appeal of what is being attempted. This observation applies equally to what is done by regional countries (which must be the nucleus of any new security structure) and by supportive outsiders.

Areas of NATO-Related Activity. If the politics can be got right, then NATO could be a useful vehicle for contributing to the functional aspects of a regional security structure for the Persian Gulf, especially in regard to practical steps required to help make the structure effective and—by diluting the apparent role of the United States in supporting regional efforts—perhaps more politically viable in the region.[22] Indeed, well beyond the fledging cooperative efforts under the rubric of ICI, NATO has now had nearly a decade and a half of experience in implementing its flagship PFP program which, by some accounts, is the most-important post–Cold War venture undertaken by the Alliance in the area of helping to foster practical security among countries that emerged from the wreckage of the Soviet Union, the Warsaw Pact, and Yugoslavia.[23] *There are many important lessons from PFP that could apply to a Persian Gulf–sponsored, –organized, and –conducted "PFP."* The regional countries could also consider the model of the Euro-Atlantic Partnership Council (EAPC), which was created in 1997 and designed to "provide the overarching framework for consultations among its members on a broad range of political and security-related issues, as part of a process that . . . [would] develop through practice."[24] Among other things, in theory,

[21] See Riga Summit Declaration, 2006.

[22] This would be true provided that NATO is not itself an inhibitor in the region either because of a general concern on the part of regional countries that they not be seen as engaged with the alliance or because of potential neuralgia over being more closely involved with the United States. If NATO were indeed judged to be an inhibitor, its role might have to be obscured.

[23] See Partnership for Peace: Framework Document, 1994.

[24] See Basic Document of the Euro-Atlantic Partnership Council, 1997.

each member of EAPC can use the council as a forum for considering problems that one member is having with another. This mechanism has not been much used, but it is available, and there are a number of unsettled problems that have been held over from the "frozen" period of the Cold War, the division of Europe, and the submerging of individual national identities in the Soviet Union and Yugoslavia.

There would, however, have to be significant differences between NATO's existing PFP and EAPC and equivalent partnerships for Persian Gulf countries. After all, a key goal of both NATO institutions has been to foster the democratization of partner militaries and, through them, of societies in general. This is consistent with the basic security parameters of the NATO Alliance and transatlantic relations, in general, which include a tripartite formulation: military security, democratic development, and economic advance. (The latter two are often carried out through the activities of the EU, individual countries, the private sector, and NGOs.) Given that it is unlikely that all, or even many, of the countries that would need to be included in a Persian Gulf security structure either are democracies, in any reasonable sense of the term, or are likely to become democracies within the time frame within which the security structure would need to be up and running to have much benefit in the next few years, NATO-inspired or -supported activities would have to be limited to functional, nonpolitical cooperative areas—a sort of PFP "lite." Indeed, in the invitation issued for ICI, the word *democracy* is used only once, and that to indicate that NATO is not the appropriate instrument for fostering democracy among possible ICI members.[25]

None of this means that democracy promotion should be dropped from the ambitions of any security structure as impractical, unattainable, or somehow inconsistent with the nature of Muslim societies: Indeed, there is a significant body of literature that has confounded the all-too-common view in the West that Islam and democracy are incompatible.[26] Further, as part of a long-term effort to provide security in the region in the broadest sense, efforts to foster societal transformation within countries—often termed *modernization*—are not only appropriate: They can also prove instrumental. This is not to say either that democracy promotion practiced along the lines of that aspect of PFP can find roots as easily as it did in Central Europe, with its historical, cultural, and political traditions, or that that democratization—along with other aspects of social transformation—must necessarily be part of viable security structures and relationships in the Persian Gulf, at least in the short or medium terms.

Methods for promoting democracy and other social transformation do matter, however, but with emphasis placed on what countries do for themselves. Just because

---

[25] Taking into account other international efforts to promote reform in the democracy and civil society fields in the countries of the region, NATO's offer to those countries of dialogue and cooperation will contribute to those efforts where NATO can add value. In particular, NATO could make a notable contribution in the security field because of its particular strengths and the experience it has gained with PFP and the Mediterranean Dialogue.

[26] See, in particular, S. Hunter, 2009.

efforts to impose democracy on Iraq, jump-started with military action, have so far fallen short does not mean that noncoercive support for social and political transformation—in a democratic direction—cannot have both merit and security utility. Nor should the cause be abandoned, *ab initio*, as a result of some limited experiments with one aspect of democracy building—the holding of elections (e.g., in Gaza) that have produced outcomes judged, in some parts of the West, as unacceptable. This judgment reflects, as much as anything else, a limited, Western-centric view of the process of evolving societies in the direction of representative governance. And, even in the history of the West, free and fair elections as a product of democratization processes came rather late. Free elections as part of democratization in emerging societies? Yes. As the critical touchstone? No.[27] In his speech at Cairo University in June 2009, President Obama got right the distinctions that need to be made and the clarity of the goals to be sought.[28]

Further, in regard to roles for outsiders, a security structure for the Persian Gulf needs to be seen, first and foremost, as something that is rooted in the region itself, whose principal members are regional countries, that takes account of relations among regional states (including relations that contain some level of stress or conflict), and that can only be effective if validated within the politics of regional states (or at least the principal regional states). If these conditions are fulfilled, then a role for NATO or any other *external institution*—if one is to be involved at all—will be only to provide a *supporting* role, rather than the *primary* role.[29] This distinction is critical. "Ownership," a current buzzword that connotes the acceptability of activities to their participants and their direct involvement, is, tautologously, about "owning." Indeed, if an outside organization, such as NATO—or an individual country, such as the United States—tries to substitute itself for local efforts to produce local results and benefits, then the

[27] The ultimate proof of a valid relationship between elections and democracy is when elections are seen as important as a *process* rather than for their particular *results*. Even in Western societies, this distinction is often hard to accept.

[28] In Cairo, President Obama said the following:

> No system of government can or should be imposed by one nation by any other. That does not lessen my commitment, however, to governments that reflect the will of the people. Each nation gives life to this principle in its own way, grounded in the traditions of its own people. America does not presume to know what is best for everyone, just as we would not presume to pick the outcome of a peaceful election. But I do have an unyielding belief that all people yearn for certain things: the ability to speak your mind and have a say in how you are governed, confidence in the rule of law and the equal administration of justice, government that is transparent and doesn't steal from the people, the freedom to live as you choose. These are not just American ideas. They are human rights. And that is why we will support them everywhere. Now, there is no straight line to realize this promise. But this much is clear. Governments that protect these rights are ultimately more stable, successful and secure. Suppressing ideas never succeeds in making them go away. America respects the right of all peaceful and law-abiding voices to be heard around the world, even if we disagree with them. And we will welcome all elected, peaceful governments, provided they govern with respect for all their people . . . . (Obama, 2009c)

[29] In military terminology, this is the distinction between a *supporting* and a *supported* activity or command.

outsiders might as well simply be in charge and accept the consequences, which, sooner or later, are likely to result in failure. Indeed, one quality of a security structure that has a chance of succeeding over the long term is that tutelage, however valid at the beginning, needs to give way to a different relationship: one between outside foster "parent" and inside nurtured "child." This is not to say that the institution itself needs, in the Marxist formulation, to wither away; rather, to use an American formulation, the local and regional members of the security organization need, at some point, to grow beyond any need for training wheels.

Thus, for NATO or another outside institution to be most effective in its efforts to promote security, it should operate on the basis of a partnership with the regional members of the explicit or implicit security structure or, if there are formal institutions and structures, with those entities themselves. This could be a mutually productive relationship, and—not to put too fine a point on it—when dealing with Persian Gulf states, finding a way to fund NATO activities without putting added financial burdens on NATO member states—always a major concern—should not be an issue, given the wealth of almost all of these countries.[30]

The *what* that NATO or individual Western countries could add is reasonably straightforward. It could include providing training, conducting exercises, developing headquarters capabilities, and promoting what, classically, have been called *confidence-building measures* (CBMs). Needless to say, unless and until all the major states in the Persian Gulf region are willing and able to participate in a regional security structure, these CBMs may be of limited utility, although they can have value even if they are developed only on a bilateral basis or just involve a few countries. It is also important to note that it is always necessary not to put more weight on a CBM device than it can bear. But, if there can be developed a basic, shared recognition that continuing competitions for power, influence, and position can, as a rule, be pursued more productively from the perspective of all the major states in the region by working *within* the system rather than *in opposition* to it (or that a major state not participating can be penalized by the institutionalized cooperation of the others—e.g., through some form of containment, even a soft containment), then CBMs almost certainly can have a positive impact.

---

[30] The author has proposed that Persian Gulf oil-producing states be asked to provide substantial funds for reconstruction and development in Afghanistan and Pakistan, not just because doing so is merited in terms of the impact of the reconstruction and development efforts on overall regional security but also in recognition of the role that the United States plays—and is expected to play in the future—in helping to provide for the security of these states in relation to threats of terrorism, challenges from Iran, or other security concerns. See Gwertzman, 2009, quoting the author:

> One idea I have is that we should be asking the oil-producing states of the Persian Gulf to be putting up major amounts of money both for Afghanistan and for Pakistan. The equation is fairly simple. Our oil money goes to these countries. Some of that money should go to Afghanistan and Pakistan, especially because the oil-producing states of the southern littoral, the UAE and around to Saudi Arabia, expect us to take care of their security. Well, let them start to do what we need. And I'm talking about $10, $20, $30 billion.

Furthermore, NATO has now created a new command structure in which the old Atlantic Command has been replaced by Allied Command Transformation (ACT), which is co-located in Norfolk, Virginia, with U.S. Joint Forces Command (JFCOM). From 2003 to 2009, an American officer commanded both. In September 2009, with France's full reengagement in the NATO integrated military command structure, the command of ACT passed to a French officer, but clear liaison arrangements have been maintained with JFCOM, which should ensure that ACT continues to have access to planning and other assets developed by JFCOM.[31] ACT has been developing capacities to assist the allies in adapting to the conditions of modern warfare, including the interaction, cooperation, and even integration of military and nonmilitary instruments—what is now known in NATO parlance as the *Comprehensive Approach*.[32] ACT also has under its command the Joint Analysis & Lessons Learned Centre in Monsanto, Portugal, which develops best practices for use throughout the Alliance. This is just one of the NATO tools that could be made available to militaries (and civilians) in Persian Gulf countries. NATO also has other tools, including the NATO Defence College, the Joint Force Training Centre, the Joint Warfare Centre, and the NATO School in Oberammergau, Germany. Within the political tolerances of Persian Gulf states prepared to work in cooperation with NATO, either individually or as part of a broader and more-encompassing regional security structure, all of these tools could easily be made available.[33]

The *where* of NATO engagement would need to be decided largely in terms of the tolerances of participating members. One issue that has so far limited implementation of ICI is the reluctance of some Arab regional states to be seen to be too closely associated with a Western institution or, in fact, any *foreign* institution.[34] NATO has still not come to closure on where to locate a training base in the region. This base will be

---

[31] As a mark of its seriousness of purpose, as a harbinger of positive developments within ACT, and building on the leadership and success of the outgoing U.S. commander, General James Mattis, U.S. Marine Corps (who retains the JFCOM command), France has appointed its Air Force Chief of Staff, General Stéphane Abrial, to command ACT. See Ministère de la Défense, 2009. NATO's Joint Command Lisbon also passes to France, which nominated Maj. Gen. Philippe Stoltz, deputy commander of the NATO-led Kosovo Force and former chief of staff of the French Special Operations Command. See North Atlantic Treaty Organization, undated-a.

[32] This term has, in most cases, replaced the far-more-cumbersome military term *effects-based approach to operations*, which essentially means, *What are we trying to do and how should we do it?*

[33] North Atlantic Treaty Organization, undated-c.

[34] See, for example, Legrenzi, 2007:

> It is highly doubtful, however, whether NATO experience in Eastern Europe will prove a useful guide in building a partnership with the GCC countries. The approach taken so far by NATO officials in charge of the initiative makes a brave assumption, namely that these countries are eager to jettison the legacy of the past and are in favor of modernizing their security apparatuses along Western lines in the near term. However, the rulers of the GCC states adhere to an extremely gradual model of reform that is dictated endogenously. The idea that this change can be dictated, or even strongly supported, from the outside is perceived as very problematic.

Some others, however, are more optimistic. See, for instance, El Kamel, 2007:

part of the new Training Cooperation Initiative that RAND launched at the November 2006 Riga summit. It is designed to serve both ICI and Mediterranean Dialogue members and to include a Security Cooperation Centre "owned" by the regional members.[35] Morocco, with its ambitions to join the EU and to gain as much as possible from NATO's Mediterranean Dialogue, has informally offered a training base for NATO to use. Both Jordan and Qatar have been candidates for the location of the new base;[36] some of the allies have been inclined to choose the latter because Qatar would finance the base. In general, there could be value in locating some cooperative activities in Europe, out of sight of the Middle East,[37] or in locating them in regional countries. Politics would largely determine the best locations.

There is one other area in which a NATO model could be usefully applied in the Persian Gulf region. This is the work of NATO's Senior Civil Emergency Planning Committee (SCEPC), which was created during the Cold War to provide a coordinating and advisory function, primarily in the event of conflict.[38] It has reshaped its mandate in the post–Cold War era to focus not just on "possible use of chemical, biological, radiological weapons by terrorists" but also on "natural disasters, such as earthquakes or floods and man-made disasters [that] pose a serious threat to civilian populations."[39] In particular,

> [t]he SCEPC . . . coordinates and provides direction and guidance for eight specialised planning boards and committees. These bring together national government, industry experts and military representatives to coordinate emergency planning in areas such as: civil aviation; civil protection; food and agriculture; industrial production and supply; inland surface transport; medical matters; ocean shipping; civil aviation; civil electronic and postal communications. Their primary purpose is to develop procedures for use in crisis situation [sic].[40]

> [W]e do have a strong responsibility toward the future generation for achieving peace, stability, security in the region and beyond. I am very optimistic about the future contribution that the Mediterranean Dialogue and the Istanbul Cooperation Initiative could achieve toward the accomplishment of this noble goal.

[35] See North Atlantic Treaty Organization, 2006.

[36] See Hamzeh, 2007.

[37] The United States has extensive training facilities, originally expanded because of the Balkan conflicts of the 1990s, at Grafenwoehr and Hohenfels, Germany.

[38] See North Atlantic Treaty Organization, undated-b.

[39] North Atlantic Treaty Organization, undated-b.

[40] North Atlantic Treaty Organization, undated-g. Furthermore,

> [t]he SCEPC also oversees the activities of the Euro-Atlantic Disaster Response Coordination Centre . . . at NATO Headquarters, which acts as the focal point for coordinating disaster relief efforts among NATO and partner countries. All NATO member countries are represented on the SCEPC, and some meetings are also open to NATO's Partner countries. (North Atlantic Treaty Organization, undated-g)

Given that a first charge on governments, almost regardless of their organization or ideology, is to help provide relief for their populations in the event of natural or other disasters, the SCEPC model could be applicable in the Persian Gulf as one of a range of CBMs.

The European Union

The direct military engagement of the United States in Persian Gulf security—or its indirect engagement through the NATO Alliance—has one overwhelming inherent advantage in that it indicates, to potential friends and foes alike, that the United States is strategically committed to preserving its interests in the region and, by extension, the interests of its allies and partners. At the same time, there is an inherent *disadvantage*, at least in some countries and with some populations, because of the lightning-rod effect afforded to the enemies of the West and the mobilizers of anti-Western opinion, including terrorists bent on using such anti-Western or anti-U.S. attitudes as a vehicle for recruiting adherents. This disadvantage is supplemented by efforts to paint the United States as the successor of British and French colonialism in the area, a legacy the United States originally inherited through no fault of its own as it progressively replaced, from the implementation of the Truman Doctrine through Britain's abandonment of its East of Suez policy in the late 1960s, the presence and influence of the two European colonial powers. The United States has subsequently acquired the coloration for many regional observers as a neocolonial power, an image further reinforced by the U.S. invasion of Iraq in 2003, which looked, to many regional parties (peoples more so than governments), as yet another Western colonial venture.[41] These attitudes have been allied, of course, with the commitment of the United States to Israel's security and future evidenced through extensive support over many decades, which then leads the United States to be seen by many observers in the region as somehow responsible for all of Israel's actions, including in its struggle with the Palestinians.

The balance of advantage must rest on the side of supporting an extensive U.S. role in the Persian Gulf region, however it is exercised, and, to the extent that any external power is involved, with the United States' having lead responsibility for creating and managing an effective security structure, whether informal or highly elaborated. Nevertheless, there is virtue in the United States' playing less of a visible role, provided, of course, there are no misperceptions that this signals U.S. abstention from the exercise of security responsibilities when such exercise is needed.

Among Western capabilities, along with what individual European countries could do on a bilateral basis, the EU could both play a part in helping to provide security and stability in the Persian Gulf region and adopt some formal role in regard to a new regional security structure. That role could include an economic dimension, which

---

[41] Iran has sought to exploit this image of the United States for the past 30 years in, for example, formulations like the "Great Satan."

would be an extension of a long history of both informal and formal institutional involvement, dating from at least the original 1972 Global Mediterranean Policy[42] launched by the then European Community and followed up, a year later, with the Euro-Arab Dialogue inaugurated after the Arab oil embargo in 1973.[43] In November 1995, the EU began its Euro-Mediterranean Partnership (also known as the Barcelona Process), which, although formally about Mediterranean countries, also has a window into developments farther east.[44] And, in July 2008, President Sarkozy, with France then acting as the EU president, held a major conference to create a Union for the Mediterranean. This was a "relaunching" of the Barcelona Process. Again, it stopped short of countries in the Persian Gulf region,[45] but it could provide a basis for extending the Barcelona Process, or a variant of it, to the Gulf region.

This economic dimension of a potential EU role should not be underestimated, even though many of the countries in the Persian Gulf are extraordinarily wealthy, at least in terms of revenues from hydrocarbons. Not only is there the web of relationships between individual European countries, companies, and financial institutions on the one hand and regional counterparts on the other, but the EU itself has significant potential for organizing and conducting a vast array of economic relationships that, taken together, could help to promote security in the broadest sense. These activities could also be part of a shaping effort that helps to promote not just Western interests but also those of individuals and groups within regional societies that wish to make themselves proof (to the degree possible) against invidious external influences or against terrorism, from wherever it emanates.[46]

There is more that the EU can do in the region of the Persian Gulf as part of its twin ventures in organizing to conduct mature foreign and defense policies. These ventures have been divided formally into its CFSP and its European Security and Defence Policy (ESDP);[47] the latter was renamed the *Common Security and Defence Policy* (CSDP) under the Lisbon Treaty that went into effect on December 1, 2009.[48]

---

[42] See Miller and Mishrif, 2005: "[T]he Community's Global Mediterranean Policy, introduced in 1972, which concluded a series of trade agreements between the EEC [European Economic Community] and Syria, Iraq, Jordan, and Lebanon had little impact on economic development in the Arab world."

[43] Miller and Mishrif, 2005. The Euro-Arab Dialogue was suspended after the Egyptian-Israeli Peace Treaty in 1979.

[44] See Barcelona Declaration, 1995.

[45] See Joint Declaration of the Paris Summit for the Mediterranean, 2008.

[46] See Chapter Ten. The author has also proposed a U.S.-EU strategic partnership in regard to this kind of shaping effort in the areas of health, education, job creation, and the like. See R. Hunter, 2004.

[47] For an introduction to CFSP and ESDP, see R. Hunter, 2002.

[48] The Lisbon Treaty also combined the functions of the High Representative for Foreign and Security Policy (which was based in the European Council) and the European Commissioner for External Relations and European Neighbourhood Policy (which was based in the Commission) into a new High Representative of the Union for Foreign Affairs and Security Policy, who also serves as Vice President of the European Commission. This also

These institutions developed over the years as part of the overall evolution of the EU, building on efforts as old as the abortive European Defence Community of the mid-1950s. These efforts have advanced in fits and starts in recent years but have begun to take on both more of a personality of their own and increasing roles in present-day crises, following the basic framework of the so-called Petersberg Tasks, adopted by the Western European Union in June 1992.[49] CSDP incorporates a set of practical tools, including planning, headquarters, and command-structure arrangements. Among these are the Political and Security Committee, roughly analogous to NATO's North Atlantic Council (the two meet from time to time); the European Union Military Committee; the European Union Military Staff; and the Civilian Planning and Conduct Capability.[50] ESDP/CSDP has played an active role in a number of crises and longer-term engagements, notably in Bosnia, Kosovo, and the Congo.[51] It has also had some limited engagement in Afghanistan, and, in October 2008, it assumed new duties with the EU Monitoring Mission in Georgia.[52]

Whether CSDP acts alone or supports a regional security structure, the actual military capacity of the organization is unlikely to be very significant in terms of potential requirements for helping to keep the peace in the Persian Gulf region, even though the new EU battle-group concept worked out under ESDP could, in time, provide up to 1,500 soldiers for a limited period of deployment outside Europe.[53] Nor would CSDP need to be engaged, in terms of resource or fighting capacity, if NATO were engaged, except if political sensitivity in one country or another demanded it—a non-trivial factor.[54] But would the EU, as an institution, be prepared to become involved on an enduring basis in Persian Gulf security? The Middle East is, notoriously, a tar pit. At the best of times, most European allies are loath to become directly involved there, preferring to cede engagement and leadership to the United States, even though the EU is a member of the Quartet and has taken part in European negotiations with Iran over its nuclear programs. (Both activities would be either peripheral or ancillary tasks in relation to a Persian Gulf security structure, although the EU's role in nego-

---

includes a new European External Action Service, which is designed to be the EU's diplomatic corps. Some of the new relationships and roles under the Lisbon Treaty arrangements have yet to be worked out in practice. See Missiroli, 2008; General Secretariat of the Council of the EU, 2009.

[49] See Petersberg Declaration, 1992.

[50] See Council of the European Union, undated-a.

[51] It has, so far, undertaken 23 operations, of which six have involved conflict situations of one form or another. See, in particular, Solana, 2009.

[52] See Council of the European Union, undated-b.

[53] See for instance, Kinnunen, 2007, slide 9.

[54] It also needs to be understood that, except for military personnel serving in headquarters and some other specialized functions, the European troops that would be employed by CSDP are identical to those that would be employed by NATO. In fact, there is only one set of troops to be called upon by either NATO or CSDP, not two.

tiating with Iran on the nuclear issue could provide some useful expertise.) Indeed, ESDP passed up on the opportunity to play a role in Lebanon after the 2006 Israeli-Hezbollah conflict, even though it would have been a natural choice to take on some of the peacekeeping functions.[55]

But CSDP—in league with CFSP—does offer an alternative to NATO in circumstances when there could be political value in the region in activities being undertaken by an institution that does not include the United States. These activities could include training (even though, in practice, CSDP training is likely to be conducted alongside that provided by NATO). From Washington's point of view under the George W. Bush administration, there was virtue in having Solana, the High Representative for ESDP,[56] be part of European efforts to negotiate with Iran regarding its nuclear programs: The United States could keep its distance from the negotiations (whether that proved wise is another question), and the Iranians could engage with an interlocutor who, clearly, interacted with the United States but, with his British, French, and German counterparts, *was not* the United States.[57]

CSDP and the EU more generally could also act as alternatives to the United States in dealings with Persian Gulf on a wide variety of sensitive matters. Further, the EU states should be willing to undertake some responsibilities, especially in the economic realm but, potentially, also in the realm of noncombat military involvement, if only because of their concerns about the impact that Muslim migration is having on most European countries and because of the relationship between Muslim migrant integration within European societies and political developments in the Middle East (especially in North Africa but also in the zone of Arab-Israeli conflict and, to a lesser degree, East of Suez).[58]

The EU could also play an instrumental role in bridging or melding military and nonmilitary activities within the region of the Persian Gulf. Indeed, unlike NATO, which, at least so far, only becomes seized of a crisis when it reaches the point at which military action is being contemplated, the EU can begin dealing, through CFSP, with an incipient crisis from an early stage, and it can then continue acting in the political-military realm through CSDP. This can be accomplished even more easily now under

---

[55] See Kinnunen, 2007, slides 10–12.

[56] Solana, who has the advantage of having served as secretary-general of NATO (1995–1999), was dual-hatted as high representative for the CFSP and secretary-general of the Council of the European Union, which gave him a broader purview than he would have had as just the leader of the EU's security wing. He was also secretary-general of the Western European Union until that position was abolished (i.e., amalgamated into ESDP).

[57] It is likely that Iran would welcome U.S. participation in direct talks because only the United States can offer some of the benefits that Iran is seeking—prominently, security guarantees.

[58] As previously noted, a large fraction of Muslim Arab migrants to Western Europe comes from the Maghreb (North Africa) rather than the Mashreq (the Levant to the Persian Gulf), but the concern of European states about the relationship between developments in the Middle East, and, in particular the Israeli-Palestinian question, and their Muslim populations is still an important political factor in European attitudes.

the newly empowered High Representative for Foreign Affairs and Security Policy. In theory, EU mechanisms offer "one-stop shopping," although limited capacities and limited political will to act create major inhibitions on the EU's ability to play a critical role, including in the Middle East.[59] Furthermore, the EU already has capacities in nonmilitary areas, including reconstruction, some aspects of development and other civilian tasks, and the promotion of governance, that NATO could develop but only at the price of duplicating the skills of other institutions—notably, the EU and the UN, including the latter's affiliated agencies.[60]

More importantly, although the creation of a Persian Gulf "European Union" is some decades off—if it ever were to transpire—some of the processes, methods, practices, and mechanisms developed within the EU, including CFSP and CSDP, could contribute to developing a model for Persian Gulf nations that would help to meet their particular security needs. Such mechanisms could, for example, facilitate efforts to get the different states to coalesce around common positions in this realm. The EU and its predecessor institutions have developed techniques for dealing with tensions among member states. This is not to say that there is sufficient parallel in the Persian Gulf in terms of the region's character and history to permit easy adoption of the EU's functional approach to reducing tensions and engaging in what could be called *preemptive conflict resolution*, an area in which the European states have made historic achievements; nevertheless, the Europeans' experience could be of some relevance to states in the Persian Gulf region, provided that the political will exists to find a means of regulating relationships and reducing both the sources and emanations of conflict.

The Organization for Security and Cooperation in Europe: A Persian Gulf Variant?

One of the most-successful developments in European security during the latter part of the Cold War was the Conference on Security and Co-operation in Europe (CSCE), which was an outgrowth of the Helsinki Final Act of 1975.[61] As part of efforts to mitigate the effects of the Cold War and to prosecute what had come to be known as *détente* between East and West, CSCE involved a basic trade-off between a Western desire to promote human rights and the possibility of peaceful political change in the European communist countries and a Soviet desire to gain greater Western acknowledgment of the borders of these countries—in effect, a ratification of the agreements reached at the Yalta Conference in February 1945. In the event, however, the human-rights provisions, plus the Final Act's undercutting of the Soviet claim that the West

---

[59] This inherently greater flexibility on the part of the EU in dealing with the full range of aspects of a crisis, at least in theory, is one reason for promoting closer cooperative NATO-EU relations.

[60] Whether NATO should develop more nonmilitary (i.e., civilian) capacities and roles is an important area of debate within the context of developing revisions to its strategic concept, which the alliance plans to adopt in November 2010 in Lisbon, Portugal, during its next summit.

[61] See Conference on Security and Co-operation in Europe: Final Act, 1975.

actively sought to change European borders to the detriment of the security of the Warsaw Pact states, helped hasten the end of the Cold War. It proved to be an almost-unintended weapon to help hollow out the twin Soviet empires in Eastern Europe and within the Soviet Union. CSCE was, in its essence, a *process* institution; over time, the process worked.

The success of this process in Europe, which involved countries that, on the two sides of confrontation, had a greater or lesser degree of hostility toward one another, raises the question whether a similar institution—a similar process—could be applicable to the Middle East and, more particularly, to the region of the Persian Gulf. Further, since the end of the Cold War, CSCE continued to exist, although, in 1995, it was renamed the Organization for Security and Cooperation in Europe (OSCE). It has been engaged in a broad range of activities in virtually all zones of conflict or tension in Europe and beyond, involving both the original CSCE member states and new members (i.e., newly constituted independent countries grandfathered into Europe by dint of their having emerged from the wreckage of the Soviet Union and Yugoslavia and the partition of Czechoslovakia). Hence, OSCE now engages in activities as far afield from "Europe" proper as the Caucasus, especially as it tries to deal with the conflict over Nagorno-Karabakh and, in a limited fashion, with the situation in Afghanistan (an area technically beyond OSCE's geographic purview).[62]

Introducing the history of CSCE/OSCE is not to argue that this institution itself should be extended to include nations in the Middle East—an idea that would likely suit no one. It is rather to explore the idea whether an organization that draws on aspects of the experience of CSCE/OSCE could (1) be appropriate for the region in helping to build security and (2) play a role in support of a regional security structure. This idea has been in play for some time, in particular in regard to the Arab-Israeli conflict. Britain made such a proposal as long ago as 1996,[63] and such a provision was also included in the Israeli-Jordanian Peace Treaty of October 1994 (but never implemented).[64] Of course, for OSCE-type arrangements to be effective (or, more particularly, CSCE-type arrangements, as this earlier construct is more applicable, given the existence of basic disagreements, tensions, and the potential for conflict in the region), there has to be a basic set of agreements, or at least understandings, about the overall shape of security

---

[62] See, for instance, "OSCE Working to Assist Afghanistan," 2008.

[63] See, for instance, Netanyahu, 1996. Prime Minister Netanyahu said, "We are fully engaged in the Barcelona Process and we have accepted the recent British initiative for an OSCE in the Middle East."

[64] See the following from Treaty of Peace Between the State of Israel and the Hashemite Kingdom of Jordan, 1994, Article 4, Security, 1.b:

> Towards that goal the Parties recognise the achievements of the European Community and European Union in the development of the Conference on Security and Co-operation in Europe (CSCE) and commit themselves to the creation, in the Middle East, of a CSCME (Conference on Security and Co-operation in the Middle East). This commitment entails the adoption of regional models of security successfully implemented in the post World War era (along the lines of the Helsinki process) culminating in a regional zone of security and stability.

within a region; a prevailing desire on the part of all participants to find a means of living together, more or less peacefully; and a concomitant shared aspiration to reduce the risk of conflict.[65] These requirements clearly have not obtained throughout the zone of the Arab-Israeli conflict. Nor do they obtain everywhere in the Persian Gulf, certainly in regard to Iran on one side of the Gulf and Arab states on the other and even, from time to time, among the Arab states themselves. Indeed, the value of a CSCE-type arrangement would apply precisely when the parties had business they wanted to do together, especially in wanting to reduce the risks of open conflict and the costs of hostile relations, in circumstances when they would not be able to agree on some of the most-basic issues at stake in their relationships.[66]

Furthermore, the application of a CSCE model in the region of the Persian Gulf would also have to consider what is *not* to be replicated.[67] The Helsinki Final Act of 1975,[68] after all, was not just about "Questions Relating to Security in Europe"—the focus being explored in this work as a potential model for the Persian Gulf region—but also two other elements: "Co-operation in the Field of Economics, of Science and Technology and of the Environment" and "Co-operation in Humanitarian and Other Fields."[69] Together, these three elements came to be known as the *three baskets*. For

---

[65] The extent to which this requirement is paramount can be seen in the fact that, within OSCE, decisions have to be taken by consensus, except in cases in which action is being taken to discipline a member. In that case, that member is not allowed to forestall consensus (i.e., it is not allowed to veto action against itself). If action is being taken to get two members to resolve a dispute with one another, neither is allowed to forestall consensus. In practice, however, these situations rarely arise, and the unanimity rule effectively applies. If such a rule applied to a Persian Gulf "CSCE," there would have to be different voting rules, the organization's purview would have to be quite narrow, or the organization would have to come into being at a time when the impetus toward cooperation was clearly prevailing over tensions and potential conflict among members.

[66] This role for a CSCE-type institution can be seen as a preconflict or nonconflict process analogous to the doctrine of limited war evolved between the United States and the Soviet Union during the Cold War, when the political objectives of conflict were kept under control—e.g., in Korea and Vietnam—to reduce the chance that forms of conflict (in particular, nuclear conflict) could occur, which could not be justified in terms of any of the basic political issues at stake in their relationship. Thus, there was mutual agreement—achieved either through formal negotiations or tacit bargaining—that the methodology and extent of war should be limited because the political and other benefits to be gained by either side by pressing the conflict beyond these limits, compared with the losses each would likely suffer, would not be cost-effective.

[67] This point and the examples that follow deal, in major part, with some criticisms that have been advanced regarding the possibility of drawing on the experience of CSCE/OSCE in thinking about security arrangements for the Persian Gulf region. One such set of criticisms has been advanced by Richard Russell of the U.S. National Defense University, but his criticisms misunderstand the role played by CSCE during the Cold War—a time of continuing hostility among many of its members—and conflate CSCE with a collective security institution, which is not at all the focus of this work. Among other things, the idea is to build only on security issues—i.e., Basket One—not on the aspects of CSCE that helped to "hollow out" the Soviet empires. See Russell, 2005.

[68] See Conference on Security and Co-operation in Europe: Final Act, 1975.

[69] The Final Act also included a section on "Questions Relating to Security and Co-operation in the Mediterranean," but that provision never rose in importance to the status of a full basket on its own (Conference on Security and Co-operation in Europe: Final Act, 1975).

the Persian Gulf countries, Basket One is most apposite, although, with success and experience in this element, the other baskets might also come to have a role to play.[70]

A Conference on Security and Co-operation for the Persian Gulf (CSCPG), therefore, is an idea to be explored as a possibility for aiding the development of relationships, reducing tensions, and promoting détente that, in this case, would emphasize Iran's relations with its Arab neighbors in the Persian Gulf and also with Turkey, if it too were a member. A CSCPG would assume that the alternative—a resort to force to gain advantage—had already been ruled out by common, if tacit, agreement, for whatever reason (including, perhaps, memories of the huge costs of the Iran-Iraq War and awareness of the common interest of a group of essentially *rentier* states in the capacity to export hydrocarbons). In Europe, the basic reasons for supporting CSCE were that trying to change frontiers by force had become unrealistic, by all accounts; that the costs of the Cold War's continuing unabated, in terms of both money and risk, were seen by all countries to be excessive in relationship to what gains might be achieved either from continuing the *status quo* or, at even higher cost, by pursuing the alternative course of warfare; and that each party saw advantage in compromising on some principles and goals that, in the event, were less important to each of them than was the underlying principle and goal of reducing the risk of conflict. CSCE thus came into being only after many years of both tension and stasis in the overall East-West relationship in Europe. It also came several years after the strategic logic of the East-West (and, especially, U.S.-Soviet) military relationship argued against either side's attempting to achieve gains through open conflict.[71] Notably, the Helsinki Final Act

[70] Of course, as with any model, full incorporation would not be possible, and many elements would not be directly relevant to security issues in the Persian Gulf, at least narrowly defined. Notably, in Conference on Security and Co-operation in Europe: Final Act, 1975, Basket One—"Questions Relating to Security in Europe"—there is a subsection on "Respect for human rights and fundamental freedoms, including the freedom of thought, conscience, religion or belief" (VII) that, along with much of Basket Three, would be too much for some of the applicable Persian Gulf countries to accept, perhaps even as a long-term goal, at this point in their development. These countries are unlikely to subscribe to principles that could lead to the undermining of regimes, except, possibly, as an exercise in hypocrisy. Basket Three was the part of the Helsinki Final Act that did most to legitimate efforts to "hollow out" the Soviet Union and Warsaw Pact states, even though these provisions proved, in practice, to be at variance with two other provisions: "III. Inviolability of Frontiers," and "IV. Territorial Integrity of States." Indeed, it was this tension—the core bargain between East and West—that helped to end the Cold War and promote the dissolution of the Soviet internal and external empires. The Final Act also contains a subsection on "Equal rights and self-determination of peoples" (VIII), which would not go down well with many Persian Gulf states.

[71] The threshold period during which this strategic logic began to apply was 1962–1963, when both the United States and the Soviet Union had developed a capacity for second-strike deterrence; this capacity made détente possible, beginning in about 1967.

was agreed upon only almost a decade and a half after the strategic conditions for it had come into being.

In the Persian Gulf, therefore, some time and much effort and changes in underlying relationships—and *in perceptions* of changes in underlying relationships—may have to take place before a CSCPG could play a positive role. But, maybe not, depending on the lessons taken on board by different countries in the region about the effects, in the last few decades, of circumstances in which conflict—or at least significant interstate tension—has been a regular visitor. Thus, this is a technique that should be considered actively along with efforts to determine the precise circumstances under which it could prove useful and an exploration of steps to get there so that a CSCPG could be mutually beneficial to key partners in security for the region.

The Association of Southeast Asian Nations: A Persian Gulf Variant?

Even more relevant than a variant of the CSCE for the Persian Gulf region could be the Association of Southeast Asian Nations (ASEAN), which was created in August 1967.[72] It is currently governed by the ASEAN Charter, which was signed in November 2007.[73] In addition, there is a Treaty of Amity and Cooperation in Southeast Asia, signed in February 1976, which now includes not just the 10 members of ASEAN but also 17 other signatories, including the United States (since July 2009), China, and the Russian Federation.[74] Furthermore, in 1999, ASEAN created what has come to be

---

[72] See Treaty of Amity and Cooperation in Southeast Asia, 1976.

[73] See Charter of the Association of Southeast Asian Nations, 2007. The ASEAN members are Brunei Darussalem, Cambodia, Indonesia, Lao People's Democratic Republic, Malaysia, Myanmar, the Philippines, Singapore, Thailand, and Vietnam. The secretariat is based in Jakarta, Indonesia. Papua New Guinea and Timor-Leste are observers.

[74] The 17 additional members are, in order of accession, Papua New Guinea, China, India, Japan, Pakistan, South Korea, Russia, New Zealand, Mongolia, Australia, France, East Timor, Bangladesh, Sri Lanka, North Korea, the EU, and the United States. The treaty's six fundamental principles are as follows:

> a. Mutual respect for the independence, sovereignty, equality, territorial integrity and national identity of all nations;
>
> b. The right of every State to lead its national existence free from external interference, subversion or coersion;
>
> c. Non-interference in the internal affairs of one another;
>
> d. Settlement of differences or disputes by peaceful means;
>
> e. Renunciation of the threat or use of force;
>
> f. Effective cooperation among themselves. (Treaty of Amity and Cooperation in Southeast Asia, 1976, Article 2)

For information on the accession of nonregional countries, see Second Protocol Amending the Treaty of Amity and Cooperation in Southeast Asia, 1998.

known as "ASEAN Plus Three," which provides for cooperation with China, Japan, and the Republic of Korea.[75]

In ASEAN's own words, the treaty "remains the only indigenous [Asian] regional diplomatic instrument providing a mechanism and processes for the peaceful settlement of disputes."[76] Since 1994, ASEAN has also included the ASEAN Regional Forum (ARF) designed

- "to foster constructive dialogue and consultation on political and security issues of common interest and concern"
- "to make significant contributions to efforts towards confidence-building and preventive diplomacy in the Asia-Pacific region."[77]

In 1995, the ASEAN countries signed a treaty on a Southeast Asia Nuclear Weapon-Free Zone.[78]

As part of these efforts, in January 2004, ASEAN countries held a Ministerial Meeting on Transnational Crime Plus Three,

> where the ministers adopted the concept plan to address transnational crimes in eight areas, namely terrorism, illicit drug trafficking, trafficking in persons, sea piracy, arms smuggling, money laundering, international economic crime, and cyber crime.[79]

There has even been some thinking about broadening cooperation, building on what has been done within ASEAN. For instance, in 2008, the Australian government proposed the development of an Asia-Pacific Community.[80]

---

[75] See ASEAN Secretariat, 2009:

> [A]t their 3rd ASEAN Plus Three Summit in Manila [t]he ASEAN Plus Three Leaders expressed greater resolve and confidence in further strengthening and deepening East Asia cooperation at various levels and in various areas, particularly in economic and social, political, and other fields.

[76] ASEAN Secretariat, undated.

[77] ASEAN Secretariat, undated. In July 1996, "ARF adopted the following criteria for participants: Commitment . . . Relevance . . . Gradual expansion . . . [and] Consultations" (ASEAN Secretariat, undated).

[78] This treaty came into force in March 1997. See ASEAN Secretariat, undated: "ASEAN is now negotiating with the five nuclear weapon states on the terms of their accession to the protocol which lays down their commitments under the treaty."

[79] ASEAN Secretariat, 2009.

[80] According to a July 18, 2008, speech to the Lowy Institute for International Policy in Sydney by Australian Foreign Minister Stephen Smith, this could entail

> a regional process that would for the first time span the Asia-Pacific and include the U.S., Japan, China, India, Indonesia and other states in the region; engage in the full spectrum of dialogue, cooperation an action on strategic, security, economic and political matters [and] encourage the development of a genuine and comprehensive sense of community, whose primary operating principle was cooperation. (Quoted in Ayson, 2009, p. 34)

As with CSCE, obviously, the ASEAN model could not be imported wholesale into the Persian Gulf region. But an Association of Persian Gulf Nations would have some attractions, especially because it could permit consideration of potential security problems, including problems that could arise among the members.[81] It also could promote economic relationships not just among member states but also with outsiders, especially through a version of ASEAN's Dialogue System.[82] Among other things, this system recognizes that security and economic issues can very well be related. ASEAN is also an organization that, while enunciating common purposes, does include members that do not always see eye to eye with one another on some key matters: The roles of Myanmar (a full ASEAN member) and North Korea (a Treaty of Amity and Cooperation signatory) are the most-obvious cases in point.

Several aspects of ASEAN speak for its possible relevance to the Persian Gulf region. First, along with its appendages, ASEAN has been a long time in developing, as overall political, economic, and security relations have improved (at least among most of the members of the Treaty of Amity and Cooperation). Second, it includes mechanisms for involving external countries with interests in the region, mechanisms that were, again, developed over time. Third, not all issues have to be resolved among member states (at least for accession to the Treaty of Amity and Cooperation) before any engagement can be contemplated. Fourth, there is a generally shared vision of the value of promoting security in the broadest sense of the term.

Indeed, whether the Persian Gulf states would think in terms of an Association of Persian Gulf Nations or some other model (including a sui generis model), one interesting area of ASEAN activity could have broad application in the development of patterns of security and cooperation: economic relations among regional states. At one level, this economic cooperation already exists in OPEC, whose members include most

---

[81] Notably, there is no formal mechanism or process at NATO for dealing with disputes between full members (although, in theory, but not in practice, the Euro-Atlantic Partnership Council might be stretched to fill that role). When such disputes do occur and are relevant to NATO—e.g., in the case of Greek-Turkish disputes or the long-standing British-Spanish dispute over Gibraltar (now muted, but with both sides continuing to assert their respective positions)—any diplomatic effort at the level of the Alliance is conducted informally and on an ad hoc basis and is usually led by the secretary-general. In the 1990s, for example, there was an informal "group of five" NATO ambassadors, including the author, who assisted the secretary-general on the Greek-Turkish issue. As part of the accession-treaty ratification in the U.S. Senate for the first three Central European aspirants—Poland, Hungary, and the Czech Republic—Senator Kay Bailey Hutchinson (R-Tex.) did try to get such a mechanism created at NATO, looking toward the possibility of residual disputes of a serious nature among new NATO allies, but the Clinton administration rejected her proposed amendment. On Gibraltar, see "Spain and Britain Agree on Closer Cooperation after Historic Gibraltar Visit," 2009.

[82] "At the Second Summit [in August 1977] in Kuala Lumpur, the ASEAN heads of government agreed that the association's economic relations with other countries or groups of countries needed to be expanded and intensified" (ASEAN Secretariat, 2009). Initial outside members of the Dialogue System were Australia, Canada, the EU, Japan, New Zealand, and the United States. Since then, China, India, South Korea, Russia, and the UN Development Program have joined.

of the Persian Gulf states that would have a role in a regional security structure.[83] The fact that OPEC has continued to function effectively through all of the region's political and security perturbations is an indication of the power of at least one common interest: the export of oil and the management of a cartel (or quasicartel) to maximize revenues for all the organization's members. Further, GCC members did create a Gulf Common Market at the beginning of 2008,[84] and the GCC could, at some point, include Iran as a member—following the first-ever attendance of an Iranian president at a GCC summit meeting—and also develop ties to countries beyond the region.[85] Not surprisingly, there is a long way between deciding to create a Gulf Common Market and its practical realization. From the standpoint of this work, the important point is that at least six key countries in the Persian Gulf—countries that are members of the GCC—have moved in this direction and even considered a place for Iran. This could bolster the concept of a regional security structure, not just in terms of some common economic interests (in addition to hydrocarbons) but, more importantly, in terms of the political possibilities of cooperation. Of course, there will be much to be achieved, as witnessed, for instance, by the continuing disagreement between the GCC and Iran over three Persian Gulf islands (Abu Musa and the Greater and Lesser Tunbs), an issue of symbolic and substantive significance.

The Organisation of the Islamic Conference

One organization that might, under the right circumstances, be able to play a role in security-building within the region of the Persian Gulf is the Organisation of the Islamic Conference (OIC). This is not a possibility that readily springs to mind, however, because the OIC has not, historically, been a security or security-related organization, although its charter does contain a section devoted to the Peaceful Settlement

---

[83] OPEC members from the Persian Gulf region are Iran, Iraq, Kuwait, Qatar, Saudi Arabia, and the United Arab Emirates. Bahrain and Oman are not members. See Organization of the Petroleum Exporting Countries, undated.

[84] See "Text of Final Communique [sic] of the 28th GCC Summit," 2007; Wheeler, 2008.

[85] See "PGCC to Talk with Iran on Establishment of Common Market," 2008:

> Saudi Arabia's official News Agency . . . [quoted] Secretary General of PGCC [Persian Gulf Cooperation Council] Abdul Rahman bin Hamad al-Attiyah on Saturday [as] saying, "With the aim of consolidating economic ties, increasing trade exchange and creating joint investments opportunities with countries like Iran and South Korea, the PGCC will start talks with them."

Note that the Islamic Republic News Agency added the word "Persian" to the appellation "GCC." Also see "Gulf States Urge Peace with Iran," 2007. The GCC has also reiterated its claims with regard to the islands of Abu Musa and the Greater and Lesser Tunbs, occupied and controlled by Iran. See "Text of Final Communique [sic] of the 28th GCC Summit," 2007.

of Disputes (Chapter XV), including general means for doing so and the possibility of cooperation with other organizations.[86]

Furthermore, in 1999, the OIC adopted a Convention on Combating International Terrorism that is remarkably comprehensive and that is binding on all the OIC's members, including some states (for example, Iran and Syria) that cannot be said to have "clean hands."[87] However, the convention excluded the following from its definition of *terrorism*:

> Peoples struggle[,] including armed struggle against foreign occupation, aggression, colonialism, and hegemony, aimed at liberation and self-determination in accordance with the principles of international law[,] shall not be considered a terrorist crime.[88]

This provision could, in the term of art, be used to "cover a multitude of sins."[89]

Notably, all the regional countries that could logically be associated with a regional security structure—including Iran, though obviously not Israel (or Lebanon)—are members of the OIC.[90] Composition of membership does not itself suggest that the

---

[86] See Charter of the Organisation of the Islamic Conference, 2008, Chapter XV:

> Article 27: The Member States, parties to any dispute, the continuance of which may be detrimental to the interests of the Islamic Ummah or may endanger the maintenance of international peace and security, shall, seek a solution by good offices, negotiation, enquiry, mediation, conciliation, arbitration, judicial settlement or other peaceful means of their own choice. In this context good offices may include consultation with the Executive Committee and the Secretary-General.
>
> Article 28: The Organisation may cooperate with other international and regional organisations with the objective of preserving international peace and security, and settling disputes through peaceful means.

[87] Convention of the Organisation of the Islamic Conference on Combating International Terrorism, 1999.

[88] Convention of the Organisation of the Islamic Conference on Combating International Terrorism, 1999, Article 2(a).

[89] The convention specifically protects against "[a]ggression against kings and heads of state of Contracting States or against their spouses, their ascendants or descendents," and the like (Convention of the Organisation of the Islamic Conference on Combating International Terrorism, 1999, Article 2[c]). "Struggle . . . aimed at liberation and self-determination" could be taken only so far and would definitely not be tolerated at home.

[90] Regarding the development of a security system that could be seen by Israel as contributing positively to the region, the OIC's position on matters affecting Israel continues to be a significant barrier. As recently as May 2009, the 36th Session of the Council of Foreign Ministers of the OIC (May 23–25, 2009, in Damascus, Syria) adopted a new "Resolution on Islamic Office for the Boycott of Israel," which, among others things, "[s]tress[ed] the importance of upholding the Islamic boycott against Israel, as a legal means of pressure to compel Israel to abide by the resolutions of international legitimacy" (Resolution on Islamic Office for the Boycott of Israel, 2009). The OIC's potential relationship with Israel and with the Israeli-Palestinian peace process is very much part of the Obama administration's negotiating approach. Thus, the following statement by the OIC's secretary-general, Ekmeleddin Ihsanoglu, at a press conference following the May 2009 OIC Foreign Ministers' meeting,

OIC could be transformed into an organization that has a significant role in security matters, but, even so, it could provide a forum for three purposes: discussing relevant issues among all the states in the immediate region, whatever the nature and status of their relations with one another outside the context of the OIC; helping to provide legitimacy for both the process and the results of efforts to create a security structure; and fostering CBMs, certainly at the political level.[91] The role of the OIC could also become more relevant if there were an Israeli-Palestinian peace agreement followed by a positive transformation of Israel's relations (now, in most cases, nonrelations) with most if not all of the OIC members.

---

needs to be assessed both on its own terms and as a possible element in OIC bargaining with the United States: "The Arab Peace Initiative [of 2002] is a call for peace in the Middle East, and the re-establishment of the relations with Israel is possible only through its observance" (Sadikhova, 2009).

[91] See Organisation of the Islamic Conference, undated.

CHAPTER TEN

Arms Control and Confidence-Building Measures

The development of a new security structure for the Persian Gulf region can incorporate both arms-control measures and CBMs. Indeed, such measures are an essential part of a security structure that is effective because they benefit the security and political interests of the participants.[1] Arms-control measures and CBMs can have material value in providing a reference point for judging the actions of others, a benchmark for behavior, and a functional approach to building security in the largest sense of the term. The primary differences between the two types of measures can be seen in their timing, scope, and subject matter. Although the two types are certainly not mutually exclusive, arms-control agreements tend to be of a more-formal and more-encompassing nature than CBMs and also to have a major, operational significance beyond their process elements—e.g., beyond the political gains realized from concluding the arrangements and even beyond the activity of reaching the agreement itself.

Both approaches try to reduce political tensions and limit the possibility of conflict at the different levels at which conflict can take place and across the full spectrum of activities in which two or more opponents are engaged. Their history of development and usefulness is long, and they were especially prominent during the Cold War.

The basic requirement for either arms control or CBMs is mutual understanding, however attained, based on each party's calculations of its own self-interest, that reducing tensions, limiting conflict, or lowering the risk of conflict, within certain parameters, is more advantageous than risking the uncertainty of the outcome of exceeding the limits that are agreed to. In other words, if taking a particular military step—actual or incipient, as with the deployment of military forces, whether the capacity of those forces is measured in terms of their quality, quantity, or location—is likely to provoke a response that, in terms of cost-benefit calculations, makes the original step not worthwhile, there can be a mutual advantage in not moving in that direction in the first place. It is also important to minimize the risk of accident or miscalculation, which can result when something takes one or both parties into a realm or intensity of

[1] At its creation, the GCC had no mechanism for regulating relations among member states in the event of strife or conflict. The charter did provide for a Commission for the Settlement of Disputes, however, in Article Ten, but the text seems to refer just to disputes over the interpretation or implementation of the charter.

conflict where neither wishes to go and from which neither can usefully profit, either relatively or absolutely. In a system of conflictual relations in which there are more than two players, limiting the risk that an accident or miscalculation can provoke an excessive response on the part of one actor or another can also reduce the chances that a third party can provoke a conflict between the first two—a process sometimes referred to (erroneously) as *catalytic war*.[2]

The basic requirement of making arms-control agreements and CBMs work is that each of the parties to an actual, incipient, or potential conflict calculates that it is better off preventing the conflict, the accident, or the misunderstanding. But such a calculation does not necessarily mean that the two will be able to reconcile basic differences or even to avoid conflict in some other realm.[3]

A good example of accident avoidance during the Cold War was the 1972 U.S.-Soviet agreement on incidents at sea. This agreement did not in itself transform political relations between the two superpowers, but it did increase the chances that both could keep control of the dynamics of their military (in this case, naval) relationship with one another so that a limited, and perhaps accidental, scuffle would not itself lead to an escalation of tensions or conflict between the two countries that neither intended and from which both would suffer.[4]

A cardinal requirement of successful arms control and CBMs, whether undertaken through joint action and formal agreement or only through unilateral and independent (but parallel) actions is communication. As in deterrence, there is no point in putting in place both policies and the capability to carry them out if the other side is not apprised of what is being done and why. Communication can be tacit—as in terms of what and where weapons are deployed or not deployed or in how weapons are sheltered against preemptive attack—provided that both sides are able to determine what is being done through either independent means or on-site inspections.[5] But it is far better for communication to be open and explicit, even if it is conducted privately

---

[2] This is a misnomer. The essence of catalysis is that the catalytic agent emerges from the process unchanged. That is rarely the case in so-called catalytic war, where the conflict-provoking party is often swept up in the fighting.

[3] It has been noted elsewhere that one reason that Britain and Russia fought in the Crimea (1853–1856) was to avoid coming to blows in a part of the world with potentially greater consequences or where the stakes involved could create political pressures that would lead one or both of the parties to press for advantage that it would not really need but at significant and excessive cost to one or perhaps both parties.

[4] See Agreement Between the Government of the United States of America and the Government of the Union of Soviet Socialist Republics on the Prevention of Incidents on and over the High Seas, 1972.

[5] Thus, a major aspect of developing a relationship between the United States and the Soviet Union during the Cold War to minimize the chances of a crisis or conflict because of misperception of the actions of the opposing side was to make sure that the opposing side understood accurately what was being done to reinforce deterrence. The most important aspect of this joint behavior was tacit—and, later, explicit—agreement not to deny either side so-called national technical means of verification (i.e., spy satellites). This tacit approach was later supplemented by an agreement not to hide important developments that could be of a threatening nature and then—with the

through third parties and subject to deniability if its existence is revealed and one or the other party could thus be embarrassed politically at home. This communication is important even between hostile powers that are sworn enemies. Indeed, when there are risks of instability stemming from the quantity, quality, or location of deployed weaponry, such communication is even more important for parties operating in an environment of intense hostility but sharing a mutual recognition of loss if tensions lead to open conflict. Of course, any such communication has to rest on a foundation of credibility. For example, if assertions are made about self-abnegating behavior (e.g., "I will not put my aircraft on alert"), there must be some means for the other side to verify that assertion, and woe betide the credibility and, hence, enduring value of the process if either side cheats. Such deception can happen only once (e.g., if one side or the other wants to start a war). Hence the virtue in President Reagan's dictum, "trust but verify."

In the region of the Persian Gulf, therefore, individual practical efforts at arms control or CBMs, some examples of which are discussed in this chapter, are of significant importance and would need to be part of any new security structure. Such efforts include exchanges of information in formal or informal settings; the use of spy satellites; tacit permission for overflights (including drones); meetings between diplomats and military officers at international conferences; and even tolerance of spying by agents, within certain bounds. Communication can be achieved in many ways and, when there is an ambition to keep tensions within bounds, to prevent accidents and conflict through misperception, it is an important asset for all parties. Of course, for parties practicing or seeking to practice asymmetric warfare, including terrorism, such communication will not be possible or at least not desirable, since the ideas of uncertainty and surprise are key elements of the tactic—indeed, they are part of the essence of terrorism.[6]

The Political Value of Arms Control and CBMs

The potential impact that regulating military relationships can have on political relationships should not be underestimated. Step one, on both sides, is to keep the military dimension from dominating the politics. For example, to their mutual advantage, both sides must prevent the emergence of a balance of forces that is inherently instable, in

---

development of arms-control agreements—by various forms of inspection through, at the beginning, secure cameras and, ultimately, on-site inspections.

[6] Even in situations of so-called all out war, belligerents may have an interest in either strategic or tactical communication, whether tacit or explicit (as was the case, for example, during the Second World War, when each side in the European theater abstained from using poison gas, in part because of concerns that the other side might retaliate with similar weapons and in part because of experience with these weapons during the First World War, during which "friendly forces" also suffered casualties). During the Cold War, there was even a U.S. doctrine of intrawar nuclear deterrence that could be achieved through signaling by the nonuse of certain weapons or the sparing of certain targets. It is also rare that belligerents sever all forms of communication with one another.

the sense that one side or the other—or both—have to be so concerned about what might happen should a conflict begin that they are unable, psychologically or politically, to focus on the possibilities of mitigating the underlying conflict itself. This is particularly true when WMDs are involved because (1) a significant amount of damage can be caused from a single military action or limited range of military actions; (2) the side that strikes first can gain a significant or even decisive advantage (e.g., through a disarming first strike); and (3) the impact on national politics could be such that the mere existence of the other side's capacity for a WMD attack could inhibit or altogether prevent efforts to deal with the political elements of conflict.[7] During the Cold War, the United States and the Soviet Union were only able to begin ameliorating the political dimensions of their conflict after both sides developed so-called second-strike deterrence—that is, the weapons and doctrines that would permit either country to sustain a first strike with nuclear weapons and still be able to retaliate with devastating effect (causing "unacceptable damage").[8] Hence, in the Middle East, the importance of issues related to nuclear weapons, especially in the hands of countries that have not convinced their neighbors that these weapons would be used solely for deterrence or as a last resort.[9]

At the moment, the most-important issue relates to the possibility that Iran will develop nuclear weapons or at least proceed in its nuclear programs to the point that neighbors and others cannot be confident that it has *not* developed a nuclear-weapon capability, however limited. Indeed, this latter point is one of the most-important problems and one that Iran needs to bear clearly in mind. It might see itself gaining added security or prestige from becoming a nuclear-weapon state—in actuality or perception—but that would also impose costs. Among other factors, these costs could be denominated in terms of Iran's becoming a pariah state to a degree far beyond what it experiences today; seeing the coalescence of other regional states against it and their seeking the protection of the United States; possibly provoking nuclear proliferation within the region; and, thus, very likely finding that it had acquired a white elephant.[10]

---

[7] This is sometimes called an *existential threat*. For instance, the Israeli government at times argues that Iran's possession of nuclear weapons, even if in the absence of any (evident) intention of ever using them, would pose a threat of high magnitude and, hence, be unacceptable, if only because the use of even one such weapon against Israel could cause cataclysmic damage that Israel could not, in any circumstances, risk happening. By this logic, a potential enemy's mere possession of a weapon can become a threat that needs to be dealt with decisively and, hence, would be, perforce, destabilizing.

[8] The emergence, in about 1962, of second-strike deterrence between the United States and the Soviet Union made politically possible the Limited Nuclear Test-Ban Treaty of 1963, the onset of détente in about 1966–1967, and, eventually, the peaceful end of the Cold War, although that still took 27 years to achieve.

[9] This is the point often used to argue that Israeli nuclear weapons—whose existence is now almost universally acknowledged—are not provocative, at least under today's circumstances, because, the reasoning goes, Israel would only use them if it were under mortal threat.

[10] It is possible that Libyan leader Colonel Muammar Qaddafi ended his country's nuclear-weapon program at least in part out of calculations that the costs, including political costs, of proceeding would outweigh benefits,

By contrast, it is certainly conceivable that individual countries in the Persian Gulf will progressively gain confidence either that their security concerns are diminishing or that they cannot be successfully attacked by a neighbor. Indeed, even today, the risk of a conventional-force assault's occurring is, most likely, far lower than the psychology of apprehension would lead one to presume. If so, regional states could become increasingly more willing, over time, to think seriously and collectively about developing political relations that will reduce the overall levels of hostility and distrust, if not of one another's ambitions and motives, then at least of one another's ability to do something militarily about these ambitions (whether directly or, for example, through subversion). Developing such confidence will not be easy, however, if there is a significant proliferation of high-capacity weaponry within the region. The strategic nuclear relationship between the United States and the Soviet Union during the Cold War involved weapons that—with the possible exception of submarine-launched ballistic or cruise missiles—would have been launched from significant distances.[11] All would have caused great damage, and that could have led to retaliation with other weapons that could also have caused great damage. There was shared awareness of the stabilizing effect of the doctrine of mutually assured destruction.

The military stability that was built up in the Cold War between the two heavily armed adversaries—both of which understood the impact of armaments on stability and acted, separately and together, to reduce risks of unwanted conflict—is unlikely to be replicated in the Persian Gulf. Indeed, even at lesser levels of potential lethality than that associated with nuclear weapons (or other WMDs), the risk that conflict can arise through miscalculation has to be taken into account. In a region like the Persian Gulf, distances are relatively short, airborne-delivery vehicles are quite advanced (certainly in terms of speed), and the potential of a disarming first strike against a conventional-response force can be significant (as was witnessed as long ago as the Six-Day War of June 1967, when Israel crippled the Egyptian Air Force in the first few hours of combat). In such circumstances, putting a premium on acquiring survivable weapon systems, acquiring early-warning capacities (i.e., which allow for warning and response times greater than the time required for the enemy to mount an attack), and forswearing weapons that can increase uncertainties on the other side become important.

---

especially given that Libya was not at significant risk of attack from a neighbor, at least not in circumstances under which a nuclear weapon was likely to act as a deterrent. In addition, he may have calculated that becoming a nuclear-weapon state would definitely bring disadvantages in terms of relations with the outside world. Indeed, upon renouncing nuclear weapons, Colonel Qaddafi found that his personal acceptability to the West went up and that Libya's isolation was dramatically reduced. Timing, of course, was also important: The resolution of claims against Libya for the 1988 terrorism bombing of PanAm flight 103 over Lockerbie, Scotland, had just occurred.

[11] The strategic problem was heightened by the development of weapons that could be launched by submarines off the U.S. or Soviet shores and by the possibility of attacks using so-called depressed-trajectory missiles, both of which reduced the warning time of an attack. In both the United States and the Soviet Union, this development made the need for secure second-strike deterrent capabilities more urgent.

Indeed, in the Persian Gulf, the capacities of modern arms tend to outstrip the physical space within which the arms maneuver. Weapons developed in a Cold War context for use over significant distances—along with high-performance weapons developed by the leading military powers since the end of the Cold War—can become destabilizing when merely deployed by local powers in a region of short distances, limited warning time of weapon use, and significant tensions between countries.[12]

Political and Military Commissions

Enter CBMs and, as a more-formal structure, arms control.[13] To begin with, as part of a Persian Gulf security structure, there would be significant value in creating both political and military commissions, preferably multilateral, whether they included outside powers, such as the United States and European states, or were limited just to regional powers. There are precedents for such activities. One of the most notable is the Standing Consultative Commission (SCC) created by the United States and the Soviet Union as part of the 1972 treaty limiting anti–ballistic-missile systems. Even though the two powers were still locked in Cold War confrontation, they were able to negotiate arms-control treaties, including that on anti–ballistic-missile systems. Given the complexity of the issues, the chance of misunderstanding and misperception, and the potential consequences if a misunderstanding or misperception went unaddressed, they saw the importance of direct communication designed precisely to deal with such issues.[14] The SCC proved its worth.

Another element was critical to the success of U.S.-Soviet arms-control agreements and was directly related to the role of the SCC: Both sides agreed on the importance of clarity—and accuracy—in understanding both the quality and the quantity of the nuclear-weapon–related armaments that each fielded. In the absence of direct

---

[12] The development of high-performance aircraft and nonnuclear missiles for Cold War purposes, together with the subsequent availability of such weapons in countries in the Middle East, with its shorter distances between enemy states, has had a significant impact on the decreasing inherent stability of military relationships in the region.

[13] These techniques can be pursued on their own or in league with other efforts. In 2007, for example, Iranian Defense Minister Mostafa Mohammad Najjar said that "the expansion of defense cooperation between Iran and other Muslim countries is one of his ministry's priorities for the year to March 2008." Furthermore, he said, in an apparent reference to the West, "Today the enemies of Muslim unity and strength have resorted to sectarian provocations to weaken the religious identity and solidarity among Muslims" (quoted in "Defense Ties with Muslim States Underlined," 2007).

[14] See Memorandum of Understanding Between the Government of the United States of America and the Government of the Union of Soviet Socialist Republics Regarding the Establishment of a Standing Consultative Commission, 1972. Not all the legitimate allegations of violation pertained to Soviet actions. Thus, at one point, the United States covered some missile silos under construction to protect them against bad weather. This was a technical violation and, when it was questioned by the Soviet Union at the SCC, the United States removed the covers. Building trust over time, even between adversaries, was the most-important goal.

observation of one another's activities "on the ground," they devised means whereby each could discover the true state of the nuclear balance (in the most-critical areas) without physical intrusion in one another's country. Thus, among other measures, both agreed on the inviolability of "national technical means" of verification, which focused primarily on reconnaissance satellites.[15]

These methods were most applicable and relevant when nuclear weapons, which were far more powerful than any conventional weapons, where involved: Miscalculations or misunderstandings about the nuclear balance (especially about the "stability" provided by second-strike deterrence) could have been cataclysmic for everyone. Not so for determining balances in conventional weapons or for seeking to halt all testing of nuclear weapons. In both of these cases, some direct observation and on-site inspections remained important and tended to bedevil related arms-control efforts.[16] In the region of the Persian Gulf, the organization most relevant to this discussion is the International Atomic Energy Agency (IAEA), which has been deeply engaged in issues related to Iran's nuclear program, amidst considerable controversy.[17] Its inspection regimes are the most-comprehensive and are applied in "more than 145 States around the world," and it is the verification authority for the NPT.[18]

Political and military commissions established as part of a Persian Gulf security structure could develop techniques for fostering stability or, to put it another way, for requiring countries that take part in such procedures to elect to start a conflict by

---

[15] See Interim Agreement Between the United States of America and the Union of Soviet Socialist Republics on Certain Measures with Respect to the Limitation of Strategic Offensive Arms, 1972, Article V:

> 1. For the purpose of providing assurance of compliance with the provisions of this Interim Agreement, each Party shall use national technical means of verification at its disposal in a manner consistent with generally recognized principles of international law.
>
> 2. Each Party undertakes not to interfere with the national technical means of verification of the other Party operating in accordance with paragraph 1 of this Article.
>
> 3. Each Party undertakes not to use deliberate concealment measures which impede verification by national technical means of compliance with the provisions of this Interim Agreement. This obligation shall not require changes in current construction, assembly, conversion, or overhaul practices.

For a history of such efforts, see, for example, Graham and Hansen, 2007. Ensuring that national technical means were preserved was a main purpose of the SCC.

[16] See, for instance, Palmowski, 2004; Harahan and Kuhn, 1996; CTBTO Preparatory Commission, 2008.

[17] See International Atomic Energy Agency, 2003–2009.

[18] See the following from the International Atomic Energy Agency, 2003–2009:

> The IAEA is the world's nuclear inspectorate, with more than four decades of verification experience. Inspectors work to verify that safeguarded nuclear material and activities are not used for military purposes. The IAEA inspects nuclear and related facilities under safeguards agreements with more than 145 States around the world. Most agreements are with States that have internationally committed themselves not to possess nuclear weapons. These agreements are concluded pursuant to the global Treaty on the Non-Proliferation of Nuclear Weapons (NPT), for which the IAEA is the verification authority. The IAEA Department of Safeguards is the organizational hub for the IAEA's safeguards work.

choice (whether well or badly conceived) rather than through some accident or misunderstanding. This is not just an idle exercise. History is replete with examples in which situations became unstable and a country decided to preempt an attack based on faulty information about not just the intentions (which are hard to measure) but also the capabilities of a potential adversary. The two commissions suggested here can at least help to introduce more rationality into the decisionmaking process. They can also lead different parties to adjust their policies (e.g., regarding which weapons to buy and not buy and where to deploy and not deploy them). Of course, given that the military capabilities of different countries in the Middle East have notoriously not been susceptible to being calculated accurately through simple bean counting, building in tolerances is also very important.

An Incidents-at-Sea Agreement

One clear illustration of the need for taking concrete steps to prevent accidental conflict in the region of the Persian Gulf was the January 2008 incident between U.S. and Iranian naval vessels in the vicinity of the Strait of Hormuz.[19] The rights and wrongs of the dispute are less important than the fact that neither the United States nor Iran had an interest in seeing this incident escalate to a serious clash of arms or worse.[20] Indeed, the short distances, the large size of U.S. naval deployments in the region, the stakes involved for all parties (i.e., interruptions in flow of oil, the potential consequences associated with wider conflict), and the general state of tensions between the United States and Iran provide compelling arguments for negotiating an incidents-at-sea agreement, following the precedent of the 1972 U.S.-Soviet agreement. In the context of a Persian Gulf security structure, it would also be useful to extend such an agreement to all the littoral states and nonlittoral states that could have warships in the gulf. Indeed, such an agreement could precede efforts to create a formal security structure and be used to help move in that direction. See Figure 10.1 for a map of the Strait of Hormuz.

---

[19] See Jajacobs, 2008.

[20] Among other things, this incident demonstrated the risk associated with using force deployments to communicate seriousness of purpose or commitment to others' security. Careful calibration is essential and includes a sensitive understanding of what, precisely, is being communicated to the target country or countries. On March 23, 2007, Iran captured a British naval vessel and 15 sailors and Marines, who were released two weeks later. Both sides told different versions of what happened. The truth of the matter is less important than the risk that the incident could have escalated. It can be argued that the chances of such an occurrence would have been greater if a U.S. rather than British vessel had been involved. An even greater risk with naval deployments in such close quarters is that a single Iranian military unit firing a single missile into a U.S. vessel in the Persian Gulf could escalate out of all proportion. Notably, the deployment of U.S. naval vessels in the Persian Gulf was done, at least in part, to send a message. That was precisely the motive that led to the presence of the USS *Maine* in Havana harbor in February 1898. The balance of analysis argues that the *Maine* blew up because of spontaneous combustion that occurred in a coal bunker located next to a powder magazine. The result, however, was the triggering of war between the United States and Spain. See Rickover, 1976.

Figure 10.1
Map of the Strait of Hormuz

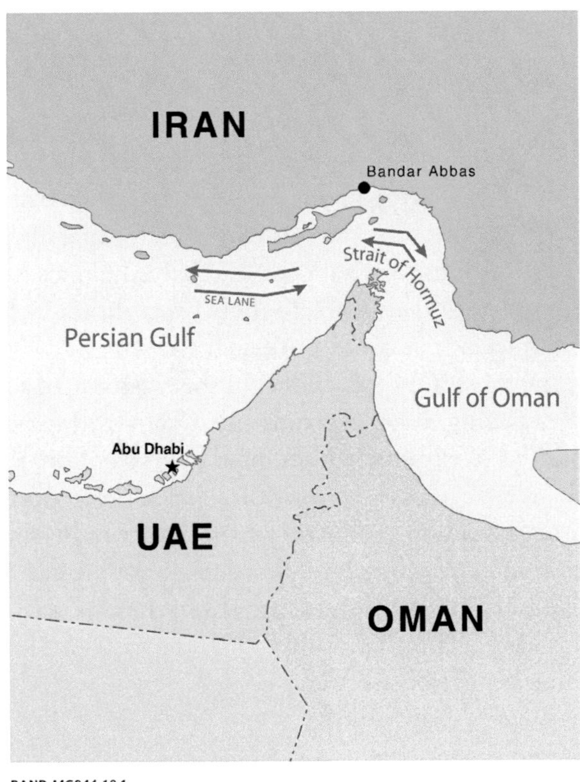

RAND *MG944-10.1*

A Freedom-of-Shipping Agreement

One of the most-important concerns of both regional and nonregional powers is the safety and security of transit routes in the region of the Persian Gulf, especially those used for oil (and, to a limited extent, natural gas) and that pass through the chokepoint that is the Strait of Hormuz. These concerns are shared by all these countries regardless of whether one or more of them might, at some point, choose to try closing the strait. Most often, the fear is about what Iran might choose to do, especially under direct threat or actual military attack. However, given the size of its economy, its dependence on oil and gas income, and its lack of any realistic alternative to exporting through the strait, Iran is as dependent on freedom of navigation—and, thus, also as vulnerable to its interruption—as any other littoral country. Iran's attempt to close the strait would, thus, almost surely occur only in response to an act of war.

At the same time, opinions vary concerning how difficult it would be to close the Persian Gulf or, more particularly, the Strait of Hormuz. The fact that the narrowest part of the strait, for the purposes of safe oil-supertanker navigation, is approximately

2 miles wide (with a 2-mile buffer zone)[21] means that the strait could be closed (more likely with mines than with sunken ships, given the depth of the waters), although it could still be difficult to achieve. In any event, the military capabilities of a state that could have the ambition to inhibit or deny transit would have to be taken closely into account. From the West's point of view, the state that could develop such a capability and the intention to use it in one circumstance or other is commonly believed to be Iran, although, theoretically, other countries (e.g., China, India, Russia) could develop such an interest and, technically, bring some elements of military or naval power to bear. Of course, there could be military counters to any such capabilities or efforts, and there would need to be calculations made about the relative balance between offensive capabilities to interdict shipping and counters to those capabilities, including strikes on either ships or missile complexes that could pose such a threat.[22]

More to the point, the most-important effect of the existence of a threat to Persian Gulf shipping could be deterring shipping companies from sailing in these waters or causing them to be subjected to extremely high insurance premiums. These two problems could be dealt with, at least in part, by providing naval protection to, for example, oil tankers and by nations' assuming the financial risk that would otherwise be borne by companies. The former method came into play during the Persian Gulf War, when the United States reflagged 11 Kuwaiti vessels engaged in the oil trade, thereby making any attack on those ships legally (and politically) an attack on a U.S. vessel. According to the theory, the threatening nation (in that case, presumed to be Iran) would have been careful not to take actions that could lead it to face combat with the U.S. Navy, four of whose ships escorted the tankers in question through the Persian Gulf.[23]

Arrangements for military (naval) protection of commercial vessels, for the potential use of military force against any source of threat to these vessels, or for assuming financial risk that were agreed upon either on a permanent or standby basis could have a deterrent effect and thus help to calm fears during a crisis. These arrangements could be part of a new security structure for the Persian Gulf if some of the key littoral states (e.g., Iran) were not taking part in the structure.[24] Such a structure would have even greater reassurance value if Iran were to decide also to become a member of the

---

[21] See "FACTBOX—The Strait of Hormuz, Iran and the Risk to Oil," 2007.

[22] In 2008, Dubai was reported to be working on plans to build a shipping canal to bypass the Strait of Hormuz, at a cost of about $200 billion. Of course, if a country could blockade the strait, it could also attack a shipping canal, thus consigning this bypass idea to the realm of fancy. See Robertson, 2008.

[23] See Armacost, 1987.

[24] This is a static analysis related to today's circumstances. It begs the question of what might transpire if there were a change of government in, say, Saudi Arabia, that brought to power a regime with a different posture regarding relations with the West or, in time, changes in Iraq such that it again became a serious power in the region. One needs to recall the Iraqi air attack on the USS *Stark* in March 1987, in which 37 Americans were killed. This was styled an accident by the Iraqis, but, at the time, the author thought it could have been a signal by President Saddam that the Persian Gulf was his "lake." As it was, the United States accepted the Iraqi explana-

freedom-of-shipping agreement, as it could very well do, given its dependence on security of supply through the strait.

In theory, all states on the Persian Gulf littoral could take part in standby arrangements regarding possible actions if freedom of shipping were under threat or some risk short of an actual attack. Such arrangements could include the sharing of financial risk (e.g., associated with insurance premiums) or even some form of cooperation to ensure freedom of navigation, beginning with a multilateral incidents-at-sea agreement. A first step would be seeking common agreement about the shared interests of all parties, both littoral states and outside countries, to guarantee the freedom of the seas in the Persian Gulf, the Arabian Sea, and the connecting Strait of Hormuz. This common agreement would include formal understandings about the limits on potentially menacing behavior on the part of all members of the agreement and about collective action against any country that posed a threat to freedom of shipping.

Counterpiracy Cooperation

The shared interest of all littoral states in freedom of the seas also provides an opportunity for mutual cooperation, which could help to build relationships that could have wider positive benefits. In recent years, there has been a significant rise in piracy, especially off the shores of East Africa and including in the approaches to the Red Sea and the Suez Canal.[25] There has even been piracy in the Persian Gulf.[26] A notable—and directly relevant—incident was the November 2008 seizure by Somali-based pirates in the Arabian Sea of the *Sirius Star*, a Saudi-owned crude-oil carrier.[27] The supertanker, with a crew of 25, was released nearly two months later after payment of a ransom reported to be about £2 million.[28] The piracy threat in general was dramatized by the April 2009 seizure of a U.S.-owned vessel, the MV *Maersk Alabama*, an incident that attracted worldwide attention, and the rescue of its captain, Richard Philips, by U.S. Navy SEALs.[29]

---

tion. Still open to speculation is whether this U.S. response helped set up President Saddam's miscalculation three years later that, if he invaded Kuwait, the United States would not respond as it eventually did.

[25] See, for example, Machefsky, 2008.

[26] See Mackay, 2007:

> The latest incident took place on August 2 [2007] when a Cyprus-flagged container ship, the MV *Sima Touba*, was attacked by armed pirates near Umm Qasr. The pirates, who were heavily armed, boarded the ship spraying gunfire, injuring the second officer. They fled after stripping the entire crew of cash and personal belongings.

[27] Glendinning and Sturcke, 2008.

[28] Pflanz, 2009.

[29] See, for instance, "Hostage Captain Rescue; Navy Snipers Kill 3 Pirates," 2009.

The implications for the security of oil and gas exports by sea from the Persian Gulf are obvious; so too is the need for all concerned nations (which include ship-owning nations and nations with an interest in the insurance industry) to take action. Naval patrolling is already being conducted by both NATO and the EU.[30] The piracy threat also provides incentives for cooperation among the Persian Gulf energy-exporting nations, both to support efforts by outsiders with military capabilities and to organize themselves for potential action. These incentives are relevant to a new Persian Gulf security structure, regardless of whether the instrumentalities of action were provided by the regional countries themselves (although their capacity to act is limited) or through their working with external actors. Here, reluctance on the part of Iran's political leadership (but, potentially, less likely on the part of its navy) to working overtly with Western countries, including the United States, or with Western institutions, such as NATO and the EU, could compete with economic self-interest in guaranteeing the reliability of supply and limiting the cost of transit (e.g., insurance costs). Indeed, the drama of recent piracy incidents is awakening the interest of actors who are becoming concerned that piracy cannot be seen simply as a cost of doing business.

This is an area in which there is a clear common interest among states both in the region and beyond, and the full range of stakeholders is involved. These include shipping owners, maritime unions, and insurance carriers. Counterpiracy cooperation can be one basis for broader, shared cooperation in a Persian Gulf region and for greater confidence in security and lessened political tensions.

A Counterterrorism Compact—and Practical Cooperation

When creating CBMs, one of the most-important efforts needs to be to gain agreement, including from all the parties to a regional security structure, about the illegitimacy of asymmetric warfare, including terrorism. This agreement should not just take the form of declarations that "terrorism is a bad thing"—there have already been quite enough of those[31]—but rather of a recognition by different countries with an overall concern about regional security that each has an interest in actually living by its word. *This acknowledgment needs to include an antiterrorism compact* that declares terrorism itself to be ille-

---

[30] See North Atlantic Treaty Organization, 2009a; EU Council Secretariat, 2009. China has also been patrolling in the Gulf of Aden, and, at the end of 2009, a Chinese admiral floated the idea of China's establishing a permanent base in the region to deal with piracy. See "China Floats Idea of First Overseas Naval Base," 2009. Such a development could have broader implications for Western interests in the region.

[31] Thus, in its September 2009 presentation to the P5 + 1 powers (the UN Security Council permanent members and Germany), Iran went to some lengths to underscore both its opposition to terrorism and its desire to cooperate in efforts to stop it. Its proposals included international cooperation on "[c]ombating common security threats by dealing effectively and firmly with the main causes of security threats including terrorism, illicit drugs, illegal migrations, organized crimes and piracy" (Islamic Republic of Iran, 2009).

gitimate and adopts and implements practical methods of cooperation that center on intelligence gathering, police work, and border controls. (The United States and other outside countries would likely be expected to provide practical assistance in countering terrorism.) The compact might build on the 1999 *Convention of the Organisation of the Islamic Conference on Combating International Terrorism*, a very comprehensive document, although some changes would need to be made in order for it to be fully effective and to gain broad acceptance.[32]

This agreement would be of a "what you don't do, I won't do" nature, similar to the self-denying ordinances that have evolved over time among major nations and that have been enshrined in the various Geneva Conventions. Indeed, no country in the region of the Persian Gulf is entirely free from problems related to what it considers to be terrorism. What is needed is political and practical means of turning that understanding into a shared respect for *others'* interests in combating terrorism.[33] This is another matter to be taken up and acted upon by the military and political commissions proposed in this work, beginning with agreement on a valid *definition of terrorism* so that differing views do not afford a means of avoiding responsibility. This is especially important in relation to Lebanon, Israel, and Palestine.

A Weapon Catalog: A Prelude to Arms Control

In a new security structure in the region, within which different parties will look for means of keeping the existence and deployment of particular kinds and amounts of high-performance weaponry from becoming, themselves, an added source of tension and, perhaps, even miscalculation or accident, there is virtue in beginning, beyond limited CBMs, a process of more-formal arms control. Part of the effort is quantitative and analytical. The Persian Gulf region is certainly overarmed in comparison with the potential threats to any of the regional countries, and it has been so for a long time.[34]

---

[32] Notably, the convention provides the following statement (also quoted in Chapter Nine), which may undercut much of its practical application, at least in regard to the Levant, as noted earlier: "Peoples struggle[,] including armed struggle against foreign occupation, aggression, colonialism, and hegemony, aimed at liberation and self-determination in accordance with the principles of international law[,] shall not be considered a terrorist crime" (Convention of the Organisation of the Islamic Conference on Combating International Terrorism, 1999).

[33] One problem will be gaining agreement, in fact and not just in theory, that the practice of terrorism against or within third countries—i.e., terrorism as understood by commonly recognized standards—also needs to cease. This is a much more difficult issue. It is one thing to gain agreement among near neighbors that they have a shared interest, underscored by the demands of deterrence and self-denying ordinances, not to foul their own regional nest. It is quite another thing, however, to get Iran to stop its activities in the Levant that meet a commonly recognized definition of terrorism, or to get the PKK to do likewise in regard to its Turkish targets in Europe and elsewhere, just to name two examples. All the Levantine territories need to be included in efforts to declare terrorism to be illegitimate and to take action on that declaration.

[34] See Kennedy, 1975.

Regional countries have wanted advanced armaments in significant quantities, and many of them have sizable oil revenues to pay for them. Arms-supplying countries have also been interested in making weaponry available, but only in part to reassure regional countries of their ability to defend themselves (against, for example, an Iranian conventional military attack on Gulf Arab states).[35] Indeed, there has long been a prevailing assumption that, to be effective, regional Arab militaries in the Persian Gulf require at least a degree of stiffening by some external major power, especially the United States.[36]

In that circumstance, arms sales to local countries would be designed more to provide deterrence than to be employed in combat. For supplier states, arms sales to the region are also a significant revenue earner and contributor to the balance of payments, helping to soak up at least some small amount of petrodollars.

Currently, at least in widespread public discussion, there is little systematic appreciation of the relationship between the military capabilities of each of the different parties in the Persian Gulf region in terms of what might transpire in actual combat among any of them.[37] Lacking that appreciation, it is difficult to calculate whether particular weapons in one or another country reduce the likelihood of conflict by accident or miscalculation by creating a reasonably stable balance of conventional weaponry or actually contribute to instability, to say nothing of potential bipolar or multipolar arms races. Even more problematically, it is almost impossible to judge the relationship between the weapon inventories—or to conduct assessments of organization, tactics, and training—of countries that have so little history of combat that could give indications of relative military effectiveness.[38]

Further, although extrapolating from the past may not be a good indicator, the course of Arab-Israeli conflicts over the years has raised serious questions about the capacity of Arab militaries to launch and sustain effective combat operations. Even when there is a history—as in sustained Iraqi combat operations against Iran—this

---

[35] This is an issue of perception that could only be validated in the actual event of hostilities. It also begs the question whether Iran (in this instance) would be interested in using military force against a neighbor beyond, say, bolstering Shi'ite groups contending for power in Iraq—itself a matter of conjecture.

[36] This idea immediately gained added currency when Secretary Clinton suggested in July 2009 that, in the event that Iran acquires a nuclear weapon, the United States could extend a "defense umbrella" over the region. See "U.S. 'Will Repel Nuclear Hopefuls,'" 2009.

[37] Some military analysts do work and publish on similar questions. See, for instance, Cordesman and Al-Rodhan, 2006.

[38] The two seeming exceptions to this generalization—Iraq's invasion of Kuwait (1990) and the Iran-Iraq War (1980–1988)—are not good examples to use when making this kind of calculation. In the former case, Iraq's military capabilities were so far superior to Kuwait's that the issue of a stable balance did not arise; in the latter case, the rupture of the Iraqi military establishment and state apparatus in 2003 and after reduces the validity of any contemporary inferences about stable or instable balances based on Iran's and Iraq's relative military capabilities at the start of their war. Also, although it cannot be proved, it is reasonable to assume from the level of casualties and other destruction in the Iran-Iraq War that neither country would choose to engage in another conflict with one another if doing so could be avoided.

does not mean that this history can be relied on as a reliable indicator of future battlefield performance, especially when the enemy in a future conflict has significantly different weapons, personnel, and tactics: Witness the almost ludicrous predications made by many Western commentators about how well the Iraqi military would acquit itself during the 1991 war and even, with less commentator hubris, during the 2003 war.[39] However unfair they may be, such collective memories render calculations about any military balance highly suspect or, at the very least, unreliable, certainly for the purpose of judging "stability" or the lack thereof: Balances are shadows at best, even if, in the event, Arab or other militaries performed up to their standards on paper. Indeed, the fog of war seems to have as great a role in the Middle East as elsewhere—perhaps greater.[40]

At the very least, therefore, the development of a new security structure for the region must include this sort of systematic analysis, which should be based both on calculations of capabilities and on plausible scenarios of conflict, in an effort to try understanding the potential role of weaponry in conflict, in short- and long-term crisis management, and in deterrence. It is to be assumed that this analysis is conducted routinely by the U.S. military and intelligence communities; also, raw figures can be found in public sources, such as *The Military Balance*, which is published by the International Institute for Strategic Studies.[41] But, for balances to be calculated and to have some stabilizing effect within the Persian Gulf region, there would need to be widespread acceptance of at least a basic framework of analysis among the key parties.[42]

There also need to be provisions for at least some form of verification of one another's military stockpiles. Ideally, this verification would include some regime for on-site inspections, perhaps administered by a third party—a model used by the IAEA in nuclear matters. This verification task could be entrusted to the military and political commissions proposed in this work. There are three basic tasks: gaining information; promoting trust in the information that becomes available (and the means whereby it becomes available); and analyzing the information, especially as regards the stability (or lack thereof) provided by the balance of relevant weaponries held by different countries. These three tasks represent a tall order and, to be truly effective, the middle task—promoting trust—must develop at a pace rapid enough so that the entire effort has a chance to succeed. This pace is difficult to achieve even among countries with relatively close relationships, and, in such an area as the Persian Gulf, where mutual suspicion is likely to be more the rule than the exception, it could prove still harder. The recurring problems associated with the role of the IAEA in monitoring the Iranian

---

[39] See, for instance, O'Hanlon, 2003.

[40] On paper, French military forces should have been more than a match for German forces in 1940.

[41] See International Institute for Strategic Studies, 2009.

[42] There is also the problem that asymmetric warfare, whether classic or terrorist-based, distorts military-balance calculations.

nuclear program and gaining support for its findings among outside, skeptical governments (some with axes to grind) illustrate this area of difficulty. (However, regarding the IAEA and Iran, the polarization of views is not primarily pronounced between different countries in the immediate region of the Persian Gulf but rather between Iran and outside countries with a declared interest in the Iranian nuclear program, especially Israel, the United States, and various European countries.)

Thus, the focus in terms of Persian Gulf conventional weaponry and the means for employing it needs to be placed to a great extent on defining, in relative terms, what matters and what does not in regard to either promoting stability between countries or undermining it. Of course, making these calculations is not just a technical exercise: It also requires assessing the psychological impact of weaponry in terms of either confidence-building or confidence-eroding effects. Making the latter, psychological calculations is more difficult than simply counting weapons. But, it could be accomplished more easily by centering the assessment of information—derived, perhaps, through on-site inspections or other means of verification—on truly important quantities and categories. In any event, this process poses challenges: Witness, for example, the difficulties experienced in trying to achieve arms-control among (relatively) willing partners during the drawn-out talks on mutual and balanced force reductions (and their successor talks), which eventually produced the Treaty on Conventional Forces in Europe. One can but try, recognizing that not trying can increase the chance that tension—or even conflict that is unintended or not welcomed by any of the involved parties—will arise.

Another aspect of analysis is also important: efforts to understand the various forms of threat—or challenge—that could be posed to the security of regional states, with *security* defined in broad terms. This is particularly important in regard to possible asymmetric warfare, a case in which classical assessments of relative military capabilities among regional states are likely to be of little utility in calculating either security requirements or the relationship between individual countries' security responses. This factor will be important in calculations about the potential role of outsiders (e.g., the United States) in helping to provide security reassurances to regional states. Indeed, absent another major military conflict in the region between local states—the Iran-Iraq War was the exception, not the rule—subversion, terrorism, and other kinds of asymmetric warfare are likely to be the key security problems to be countered. Indeed, some of the key threats to security, *as perceived by individual countries*, may not even be military in nature. Rather, they can involve migration, economic penetration, appeals to religious and ethnic minorities, and the like. This calculation about threats also speaks volumes about potential responses that could be called for, first, on the part of individual regional states; second, on the part of a collection of such states; and third—and only third—on the part of help from outsiders. Insurgency and counterinsurgency have been key preoccupations of U.S. and other Western analysts and governments in recent years.

Such a complex and compound analysis should begin with classical issues of military balances but also include the full range of potential asymmetric threats and challenges, including nonmilitary challenges to security, writ large. This analysis would also be important as a starting point for understanding the potential role of external powers (e.g., the United States) in helping to provide security—and a sense of assurance (which could also serve as deterrence)—to regional states that could not be confident of being able to defend themselves on their own against a plausible adversary. At the same time, such an understanding would help in calculations of (1) the extent to which the United States or another external state would need to have military capabilities either in the neighborhood or capable of being reintroduced rapidly and (2) what these capabilities would need to be to help preserve stability. In the final analysis, these are political judgments. Calculations made now would have to be viewed as rudimentary, especially because there is no common reference point that allows different actors to make calculations about military balances that could condition one another's approach to arms acquisitions, structure of military forces, and efforts to keep the military dimension of developments from overwhelming even enlightened efforts to limit the likelihood of open conflict or, short of that, continued tensions beyond a level that would otherwise exist in the absence of this military-weapon factor.[43] Indeed, a common appreciation could help to promote stability or, if the calculations showed that stability was at risk, guide different states—both internal and external to the region—toward polices that would help to promote stability.

Limitations on Sales and Supplies of Weapons

An analysis of potential arms balances, as difficult as it would be to undertake in this particular region, is the starting point for trying to prevent the flow of arms to the region from itself contributing to the escalation of tensions, promoting serious miscalculation in the midst of rising tensions, and, perhaps, stimulating (even unwanted) conflicts. At various times, some limits on arms sales and other arms flows into the region have been either sought or imposed by outsiders. From the United States' perspective, this has long been a factor in its calculations about which weapons to permit U.S. companies to transfer to particular states in the region and which not to allow. More often than not, however, the calculations involved do not relate to military relationships between and among regional states but rather between regional states and Israel. The United States has, indeed, been careful not to provide to Arab states weapons that might give them an advantage (or even an equalizer) in possible conflict with Israel. When the United States has wanted to provide the arms for other reasons (e.g., for confidence-building,

[43] A classic objective of arms control that was honed during the Cold War is to prevent the very existence of weapons and their patterns of deployment from increasing political tensions and, indeed, the possibility of conflict, even when neither party in the confrontation wants conflict.

for Persian Gulf security, or to gain revenue), it has been sedulous in making sure that Israel retains both its qualitative and quantitative edge in regional weaponry.

If a regionwide security structure is devised and gains broad acceptance among regional states, the effort should include, if formal arms-control agreements are not possible, at least codes of conduct about which weapons are permissible within the region and which are not; the codes should also cover weapon basing, alert status, etc. The easiest category of weapons to include in such arrangements is WMDs (whether nuclear, radiological, biological, or chemical) because of their lethality and the fact that no country can really hope to gain critical advantage from possessing them.[44] Of course, in regard to a possible Iranian nuclear-weapon capability, many other arguments and points of analysis need to be introduced, including the following: whether the introduction of such weapons in the Iranian military arsenal would provoke a nuclear or other WMD arms race in the region; how Iran might seek to exploit one or more such nuclear weapons for political purposes (e.g., intimidation) as opposed to actual military use; the risk that the mere existence of these weapons, or the possibility that they could soon be acquired, would provoke the very conflict that Iran and, presumably, other countries would like to avoid; and whether an Iranian bomb would prove to be a white elephant as opposed to a political, military, or strategic asset. Be that as it may, prudence argues both that no regional party acquire nuclear weapons and that there be concerted efforts to keep that from happening.[45] The efforts should include trying to draw Iran productively into regional security arrangements, a process that would have to entail negotiations, a search for common strategic and political ground, and some degree of mutual accommodation.

Arms control, of course, does not just mean reducing arms: It can mean *increasing* them. The emphasis is on the word *control*, and there are tight linkages to the notions of balance and, even more importantly, *stability*.[46] But, in a region where countries (and potential targets) are so close to one another and given the speed and lethality of aircraft and aircraft-borne weapons, it is hard to devise means for stability; dispersion and the hardening of aircraft and other potential key targets are possibilities. Abstention from acquiring certain kinds of capabilities also can have merit. This is an argument for great care and clear calculation in the provision of advanced weaponry to regional

---

[44] This argument's validity depends on the willingness of various parties, both inside and outside the region, to impose penalties for any use of WMDs, whatever the motivation.

[45] One element in calculations are regional views about Israel's purported nuclear-weapon capability, as judged within the context of (and in comparison with) concerns on the part of Arab states of the Persian Gulf about possible Iranian nuclear weapons and their almost-certain desire to gain added U.S. security involvement if Iran did acquire such weapons. In any event, the Israeli nuclear weapons issue is sure to reemerge from time to time.

[46] Thus, during the Cold War, to gain stability in the U.S.-Soviet nuclear balance, it was necessary for both sides to build weapons (e.g., submarine-launched ballistic missiles, silo-based ICBMs) to ensure the survival of their capabilities for nuclear deterrence.

states, action the United States has been taking in response to Iranian policies, including Iran's nuclear program and development of ballistic missiles.[47]

A further problem arises when one side or the other in potential hostilities is likely to have significant advantages in terms of high-performance weaponry. This advantage creates strong incentives for the less–militarily-capable side either to increase its own conventional combat capabilities or—especially when such an increase is not feasible—to engage in asymmetric warfare. But, as previously noted, asymmetric warfare introduces difficulties in calculating the nature of a military "balance" and, hence, of "stability."[48]

The value of trying to limit the role played by arms supply to the region of the Persian Gulf in reducing the stability of political relationships or exacerbating conflict once it starts points to an important role for outside powers: to act not just as restraining influences in the supply of weaponry or in calculating the relative performance and, hence, balance of weapons that are supplied but also to act to *offset imbalances* by providing arms as a matter of helping to promote the stability of relationships and, hence, reducing the risk of conflict. To make this technique effective in preventing conflict or the escalation of conflict, one or more outside powers would have to be willing to act as an *arms balancer*, both in advance and on an impartial basis. Doing this would be highly difficult. Among other things, there would need to be reasonably accurate assessments of arms balances, the acceptance of these assessments by local and external states—whether explicit or tacit—and the willingness of all significant arms-supplying states to subscribe to the policy. If external states did play a role in trying to offset an emerging arms imbalance—an imbalance that could be so destabilizing as to increase the risk of conflict (or the *perception* of the risk of conflict)—it would probably be important for those outside states to provide security guarantees on a *tous azimuts* basis. Yet, providing credible assurances is almost never easy to do (witness the difficulties the United States had during the Cold War in reassuring its European allies of its fealty), and this is especially so in circumstances in which the country or countries providing the assurances are not favoring one side over the other but are trying to be evenhanded in the effort to prevent conflict.

Nevertheless, a new security structure could include a joint request from the local parties to outsiders (e.g., the United States or the Europeans) that the outsiders assist

---

[47] See Chapter Five on bolstering regional defenses.

[48] With the term *asymmetric warfare* understood as unconventional military capabilities being used by a less-capable party against a more-capable party, it is possible, in theory, to devise a model for a balance of asymmetric-warfare capabilities in terms of the quality and quantity of weaponry. Other factors also come into play, however, including the political and psychological vulnerability of the contending parties. The use of terrorism by both parties to a conflict is not unknown—Germany and the Soviet Union each used it against the other during the Second World War. And, techniques that could be called either *terrorism* or *guerrilla or irregular warfare*—depending on who is defining the techniques—have been practiced by various combatants in insurgencies and counterinsurgencies (e.g., the Spanish Civil War, the Algerian War of Liberation).

in the effort to keep perceptions of imbalance in military instruments from contributing to tensions or conflict. If this occurs, it is a short step to asking the outsiders also to play an active role in helping to adjudicate disputes when either the local parties are unable to do or when international institutions (e.g., the UN, the OIC, or the International Court of Justice) are unable to do so. Certainly, arms control, when serving as an active instrument of promoting stability, requires mechanisms, in which local parties have confidence, to make the necessary assessments. The relative success of this arms-control process would be a good barometer of the seriousness with which individual parties view the whole idea of a regional security structure, which, to be most effective over the long term, must be based on a premise of impartiality—that is, the structure needs to include some version of collective *security* as opposed to just reliance on collective (or even solely national) *defense* instead.

Integrating Instruments of Power and Influence

In considering efforts to devise a new security structure for the Persian Gulf and vicinity, there is one further area for investigation. It involves a broader use of the term *security* than is common and relates to activities within the region that could help to reduce the risk of conflict and to ameliorate tensions and various potential causes of conflict in the *nonmilitary* realm. In short, one of the significant underlying causes of tension and conflict in various parts of the Middle East is the relative lack of what, for want of a better term, one can call *development* (political, social, and economic). In some cases, this lack is evident even in countries whose great national wealth from oil revenues is grossly maldistributed. This phenomenon bears serious analysis and consideration even if it is found not to have universal validity.[49]

It is now widely accepted that a great deal of recruitment into the ranks of terrorists, including many Islamist terrorist groups, derives less from an ideological or religious inspiration—which, especially at the leadership levels, is not much or at all subject to rational persuasion—than, in one way or another, from the conditions under which people live.[50] At one point, before terrorism became more prominent, this was sometimes discussed in terms of a "revolution of rising expectations," a phenomenon, seen in some countries, where progress toward economic success actually increased

---

[49] This qualification is analogous to understanding that "democratic" elections, per se, do not necessarily lead to a reduction of ambitions to dominate others or to limit either the existence or the use of instruments used to do so.

[50] In the last few decades and especially since the 9/11 terrorist attacks in the United States, a literature has been developed on all aspects of terrorism, including its origins, motivations, and recruitment patterns. It is too vast to be cited in detail here. Useful bibliographies include the following: Berry, Curtis, and Hudson, 1998; C. Reynolds, 2001; Forest, 2004; Motes, 2004; Forest et al., 2006; Armstrong, 2007.

people's tendency to take part in activities that are potentially destabilizing to societies and certainly extend beyond opposition to oppressive governments.[51]

In recent years, there has been a good deal of experience to demonstrate that producing success in situations that involve low-intensity conflict or counterinsurgency can be aided significantly by practical efforts to meld both military and nonmilitary activities. This was certainly true in both Bosnia and Kosovo during the latter half of the 1990s. In both cases, NATO-led forces operating under UN mandate kept the peace, but they were supplemented by nonmilitary efforts conducted largely under the auspices of the UN, the EU, and, when individual European and North American governments were involved, foreign and development ministries.[52] These efforts were supplemented by the activities of NGOs and representatives of the private sector.[53] Similar efforts have been undertaken in both Iraq and in Afghanistan. The latter is a particular case in point: ISAF includes forces from all 28 NATO allies plus representatives of a host of other government agencies, international institutions (notably, the UN Assistance Mission in Afghanistan), and NGOs. One instrument of note is the provincial reconstruction team, which includes military security personnel, development experts, and the capacity for building close relationships with local leaders and populations. The upshot of all of these and other nonmilitary activities is the potential to take an integrated approach to security, writ large, that is not about just the military dimension or the so-called kinetic phase of a conflict; rather, it focuses very much on nonmilitary efforts, including the so-called fourth phase of conflict (more popularly known as *nation-building*).

The efforts that are being undertaken in this vein in Iraq and Afghanistan[54] (and, now, Pakistan) will continue to be of cardinal importance in the context of those two conflicts. But the idea of integrating instruments of power and influence might also have a significant impact within the overall region of the Persian Gulf and environs. This integration could apply not just in potential conflict situations or even in military-related situations; rather, it could be apposite to a wide range of efforts that can be summarized as attempts to shape the environment—especially attempts to engage with societies for a variety of reasons, some economic and humanitarian and some that are strategic (in that success may help to reduce the risk of conflict).

Of course, for any outside country or institution to succeed at such efforts within many Middle Eastern societies is not easy to do for a variety of obvious reasons, includ-

---

[51] For a classic discussion of this topic, see Brinton, 1965.

[52] For a major presentation and analysis of these operations along with lessons learned and best practices, see R. Hunter, Gnehm, and Joulwan, 2008. It represents the conclusions and recommendations of a panel of 67 U.S, Canadian, and European senior participants that derived from deliberations conducted under the auspices of RAND and the American Academy of Diplomacy in 2006–2008.

[53] For a discussion of potential roles and limitations, see R. Hunter, Gnehm, and Joulwan, 2008.

[54] See Obama, 2009e.

ing the background of a long history of colonial and postcolonial relationships. Nor can there be a one-size-fits-all approach. Indeed, in many or most situations, the United States should not be the lead agent—in some regional countries with a colonial past, the same stricture would apply to Britain or France—and, in some cases, neither should NATO. The United Nations and its specialized agencies, which have significant expertise in these areas, may be much better placed politically than any external country, as may be a number of NGOs. The same may be true of the EU, especially with regard to operating in regional countries that can tolerate a Western, but not a U.S., role. Each situation should be approached on its own terms. But, the basic logic is sound, provided that efforts are undertaken with the appropriate appreciation of historical, cultural, religious, social, economic, and political sensitivities fully borne in mind.[55] An added value is that both European and North American countries have the capacities—some in the government and some in the private and NGO sectors—to support local efforts that can, when successful, contribute to security in the largest sense of the term. In the areas of health, education, training, job creation, and overall development, the democratic societies of Europe and North America have extraordinary capacities—especially among individual citizens and organizations—to be of instrumental value.[56]

Of course, in the region of the Persian Gulf, many of the countries involved are not challenged by a lack of resources. In fact, most of the Arab oil-exporting countries have great amounts of wealth, whether or not they choose to deploy it within their own societies and extend broad benefits to their populations. Some host sizable immigrant populations employed as servants, laborers, and construction workers. Among the Arab states of the Persian Gulf littoral, Iraq is the key exception in terms of currently having significant wealth and will remain so until the oil industry is fully recovered and further developed. Across the Persian Gulf is Iran, a country with significant hydrocarbon revenues and whose current uneven state of economic development contributes to circumstances under which nationalism and religiosity are intensified, particularly among poorer parts of the population.[57] The specific analysis discussed in this section is therefore more likely to apply to "outlying" societies in regard to the Persian Gulf and its security—societies including those of Syria, eastern Turkey, Jordan, Yemen, and, especially, Afghanistan—and to efforts to undercut the appeal of Islamist

[55] These qualifications are important. Indeed, experience has shown that there can be major limits on what outsiders, even with the best will in the world, can do in trying to shape other societies.

[56] The relationship between such engagements and promoting security was noted clearly by President George W. Bush at the International Conference on Financing for Development in Monterrey, Mexico, on March 22, 2002: "We fight against poverty because hope is an answer to terror. We fight against poverty because opportunity is a fundamental right to human dignity. We fight against poverty because faith requires it and conscience demands it" (G. W. Bush, 2002b).

[57] This point raises a question about the instrumental value of economic sanctions against Iran.

terrorism in other societies (notably, Pakistan's) that could export disruptive activity into Persian Gulf countries.

One aspect of this analysis that does have broad application in the region is the political, social, and economic relationship between government and the governed. Even without accepting the proposition that democratization can be a tool for reducing, over time, the risk of conflict, it is certainly true that issues of governance are important and are likely to rise in importance and intensity. These issues are more likely to relate to what happens within individual countries—or when these countries become more vulnerable to disruptive (e.g., terrorist) influences from outside—rather than to interstate relations, tensions, and, possibly, conflict. The exception is the seizure of a state by radical elements with an agenda of spreading a particular approach or philosophy that can be threatening to neighbors, as happened following the creation of the Islamic Republic of Iran.

Nevertheless, over time, the nature of governance in Persian Gulf societies will be a significant element in the potential for a security structure to reduce the risk of conflict and then to be able to do something about the risk. Dilemmas posed by the potential outcome of political change have a long history, and there are few, if any, universal truths.

Approaches to integrating—or at least relating—instruments of power and influence could, in time, lead to the possibility, although clearly not the certainty, of reducing not just the risk of conflict but also the requirements for the United States or other outside powers to deploy military forces in the region, including forces deployed for counterinsurgency.[58] Of course, this cannot be a straight-line process with an inertia of its own. Indeed, opponents of stability in the region (primarily, insurgents and terrorists) can be expected to work actively against these efforts, just as they have been doing both in Iraq and Afghanistan. But, in both cases, the U.S. and allied militaries, along with their civilian partners, have learned that the military instrument is, to a significant degree, the shield and that the nonmilitary efforts are the swords. As discussed earlier with regard to asymmetric warfare, this combination of instruments of power and influence can be an asymmetric remedy or a push-back effort employed by the United States, other outside states, and their local allies in the effort to appeal to hearts and minds. Indeed, as has become apparent in Afghanistan to virtually all observers, promoting better governance, reconstruction, and development is an essential part of promoting security.[59] Lessons learned and best practices developed during these efforts may be relevant in other parts of the Greater Middle East.

---

[58] In general, the Arab-Israeli conflict has resisted efforts to reduce tensions and promote peace through functional approaches in the economic field. For a major set of proposals in this area, see R. Hunter and Jones, 2006; RAND Palestinian State Study Team, 2007a; RAND Palestinian State Study Team, 2007b.

[59] See, for instance, Jones, 2008.

This nonmilitary work also offers two other benefits. Within the NATO Alliance, one of the key concerns at the moment is the question of risk- and burden-sharing, especially in Afghanistan, where NATO has lead responsibility for security but where the United States provides the lion's share of the combat forces. Many of the European allies that impose caveats on where and how their forces can be used will not change those limitations, and the United States and the other allies that are most directly engaged in combat (e.g., Britain, Canada, and the Netherlands) will continue to press them to do so. There is a great need for nonmilitary instruments, especially reconstruction and development. The instruments to implement these efforts are relatively inexpensive when compared with the cost of military deployments. and they require expertise and experience that many European countries have developed over many decades. Thus, it should be possible to see NATO Alliance burden-sharing in a broader context. In other words, allies that impose military caveats can be asked to provide significantly greater amounts of nonmilitary assistance to compensate for those limits and to participate fully in burden-sharing. In addition, the EU itself should be taking more leadership in such efforts, beginning with the appointment of a senior-level official to coordinate aid and other forms of assistance to Afghanistan (and Pakistan).[60]

This kind of arrangement in fashioning and applying nonmilitary instruments of power and influence could also be applied to other countries in and around the region of the Persian Gulf, beginning with Iraq, to help promote the goals presented in this work. Furthermore, there is an opportunity to secure needed added financing. At the moment, the United States is expected by many or most of the Persian Gulf Arab states to provide security for them against what they perceive to be external or internal threats (e.g., Iran, Al Qaeda, and its acolytes or a psychological sense of insecurity). Given that the United States and other Western countries are importing vast quantities of oil from Persian Gulf countries, thus shifting huge quantities of foreign exchange to the region, it is not unreasonable to expect the richer of the regional oil-producing countries to make a material contribution to their own security by financing a major part of the reconstruction and development work that is needed to help create lasting security in Iraq, Afghanistan, Pakistan, and, potentially, elsewhere.

In sum, an encompassing view of security in the Persian Gulf region, whose range of concerns extends beyond "classic" security issues, and of potential remedies that extend well beyond the military will help make possible the development of a new security structure for the region that brings all of these elements into the same place where they can become effective tools both for promoting overall regional security and for reducing the burdens placed on the United States and other external powers.

---

[60] See, for instance, Gwertzman, 2009.

CHAPTER ELEVEN

Conclusions and Recommendations

This work has sought to look not just at current events and immediate security requirements in the Persian Gulf region, especially with the drawdown of U.S. forces from Iraq and the continuation of tensions between Iran and a number of Western countries, but also at the longer-term future. In doing so, it has outlined and analyzed key elements of a possible new security *structure* for the Persian Gulf and environs with the twin primary goals of increasing the likelihood of long-term stability in the region and reducing requirements, over time, for U.S. and other Western engagement (compared with what otherwise would be required). It may be that the second goal will prove incompatible with the first, but this work has at least tried to put forth criteria, along with an analysis of terms and conditions, for achieving both.

This work has also focused on the regional component of security in light of changes taking place within Iraq and of increasing U.S. and allied engagement in Afghanistan and—with a much narrower set of participants—in Pakistan. At the same time, this work is not a guide to the precise policies to be followed in effectively conducting the U.S. and Coalition endgame in Iraq or in charting the strategy and tactics for the period immediately ahead in Afghanistan and Pakistan. Developments in both theaters of combat will naturally have a major impact on the regional dimensions of security. Among other things, what happens in these two areas will help to condition whether a viable security structure can be developed for the Persian Gulf and, if so, what its parameters can be and what countries can be included in it. For example, in the relatively short term, significant renewed turmoil in Iraq could require the United States to change its policies on withdrawal and would also distract from efforts to try developing a regional security structure.

One regional factor is more immediately important even than what is happening in Iraq now or may occur in the future:[1] Iran. The issue is not just what it will do (or abstain from doing) in Iraq but also its other foreign policies, especially decisions about its nuclear program. Indeed, an Iranian push toward a nuclear-weapon capability may

[1] The possibilities include a Turkish-Kurdish clash in Iraq and efforts on the part of one or more regional Arab states (e.g., Syria or Saudi Arabia) or its citizens to make more difficult the creation of a viable Iraqi government and polity.

make inevitable the development of (1) security arrangements in the Persian Gulf that involve the United States and, perhaps, other Western states in some form of long-term balancing or containment-and-deterrence policy and (2) formal security arrangements among the Arab states of the Gulf. Both of these arrangements would posit Iran as the enemy or, at least, the odd man out. Iran's behavior in several spheres, its internal troubles (including the conduct and aftermath of the 2009 presidential election), and many statements by its leadership do not help in this regard.

Important criteria for an effective Persian Gulf regional structure include the following:

- a desire on the part of a critical mass of regional states to take part in some regional security structure because they have judged that doing so is more likely to foster their security than either working against the structure or abstaining from it (and choosing instead to build security either on their own or through bilateral arrangements with other regional countries or an external actor—notably, the United States). This effort to create a security structure must include the capacity of local states to develop the necessary political will to undertake this venture, with or without outside tutelage.
- the pursuit of a building-block approach to a regional security structure, preferably prior to deciding on the merits of a regional security compact (whether collective security or collective defense)
- the willingness of member states to pursue a number of specific arms-control measures and CBMs, including
  - establishing multilateral political and military commissions to reduce tensions and the risk of conflict, including through conflict resolution
  - creating CSCE-like arrangements (i.e., a CSCPG)
  - developing PFP- and Euro-Atlantic Partnership Council–like relations between individual members of the security structure and, potentially, between them and outsiders (e.g., NATO, the EU)
  - creating a Persian Gulf counterpart to NATO's SCEPC
  - drawing on the experience of either ASEAN or the OIC in fashioning regional cooperative mechanisms
  - establishing an incidents-at-sea agreement, a freedom-of-shipping agreement, and counterpiracy cooperation; cataloguing (and defining) military capabilities; and limiting the acquisition of potentially destabilizing weapons in the region
  - adopting nonmilitary cooperation (e.g., economic relationships) that can contribute to reducing tensions and building security and integrating both military and nonmilitary approaches to security
  - creating means (and political will) to resolve or limit and contain serious intraregional disagreements, tensions, and crises

- the integration of regional security efforts within a formal UN mandate to create a rule-of-law basis for cooperation
- the creation of the security structure premised *either* on the possibility of universal membership *or* on the basis of one (or more) states being essentially "hostile" and, thus, needing to be contained rather than co-opted. These alternatives need not be mutually exclusive over time, although transitioning from the latter to the former would not be easy. This critical issue focuses, at the moment, primarily on Iran; its attitudes and policies toward other states, the region as a whole, and the outside world; and antipathy toward Iran on the part of some regional Arab states and nonregional countries.
- agreement to an antiterrorism pact[2] with "teeth" (i.e., not just a formal declaration) and opposition to asymmetric warfare, however and by whomever practiced and with practical support from the United States and other outside countries in countering terrorism in the region
- possible roles for external security-related institutions (notably, NATO and the EU and, perhaps, the ICI and broadening of the EU's Mediterranean Initiative). Such institutions could play a direct part in the security structure or simply serve as models.
- regional capacities—whether on the part of the members of the security structure on their own or in conjunction with external partners (e.g., countries, such as the United States, or institutions, such as NATO and the EU)—to deal with asymmetric threats
- roles (if any) for other external powers—notably, Russia, China, and India—that contribute to regional security rather than detract from it
- roles for other outside countries, to the extent that this is acceptable to members of a security structure, in providing support to regional efforts. This needs to include sensitivity on the part of outsiders to local concerns regarding religion, ethnicity, history, culture, etc.
- roles (if any) for forces provided by external countries (primarily the United States, but also, potentially, European nations) in accordance with arrangements to provide reassurance to some or all of the members of a Persian Gulf regional security structure. If such forces are given a role, it would be necessary to determine what kinds of forces should be included, where they should be deployed (i.e., under or over the horizon), what mechanisms could trigger their involvement, what training and other demonstrations of presence (including joint training and maneuvers with local forces) they would provide, what rules of engagement they would follow, and where and how they should *not* be based or involved. The role of arms supply in bolstering the defenses of regional countries—i.e., in the face of potential challenges from Iran—would also need to be considered and implemented in

[2] This could build on the OIC *Convention on Combating International Terrorism*.

a way that increases regional security rather than risks an uncontrollable conventional arms race. In theory, outside efforts could constitute an existential commitment—i.e., to help support regional efforts to provide security on a *collective* basis. More likely, however, it would be made on the basis of *defense*, and that implies that one or more local countries are perceived as opponents of an effective security structure. This latter situation is not preferred, but it may be unavoidable or irreversible once it comes into being.

- possible formal U.S. security guarantees of one form or another
- progressive, rather than one-shot, development of such a structure. It could be useful to hold a regional security conference (or series of such conferences) involving (1) as many regional countries as are prepared to take part and (2) outsiders, in defined capacities. The UN in particular should be involved, and other external institutions (e.g., NATO, the EU) and outside countries (notably, the United States and selected European and Asian nations) may have a place. If the United States can achieve its overall goals in the region and is not bent on playing a dominant role there for its own sake, a regional security structure based on local countries without a significant—and, certainly, without an overbearing—role for any outside power is to be preferred. That may not be possible, however, especially if some or all of the local powers argued that a sense of security underpinned by the United States (or by European powers) is a necessary condition for considering a security structure.
- long-term efforts aimed at internal social, economic, and political development, where needed. This development should be conducted both for it is own sake and as a means of both strengthening resistance to external threats and challenges (e.g., terrorism) and helping to promote regional cooperation.

Many pieces of the puzzle now exist, although many of the elements will need further refinement, and many details, including in regard to force and other requirements in and around the region to help make a regional security structure effective, have still to be considered. What has been presented here is a framework for analysis, a setting of parameters, and an invitation to further imagination, insight, and—the indispensable requirement—leadership on the part of regional states; the United States, its allies, and partners; and international institutions, in and out of the Persian Gulf.

APPENDIX

Documents

The following three documents are presented here because they are particularly relevant to analysis and proposals in this book.

Clinton Proposal on Israeli-Palestinian Peace[1]

President Clinton:

Territory:

Based on what I heard, I believe that the solution should be in the mid–90%'s, between 94–96% of the West Bank territory of the Palestinian State.

The land annexed by Israel should be compensated by a land swap of 1–3% in addition to territorial arrangements such as a permanent safe passage.

The Parties also should consider the swap of leased land to meet their respective needs. There are creative ways of doing this that should address Palestinian and Israeli needs and concerns.

The Parties should develop a map consistent with the following criteria:

- 80% of settlers in blocks.
- Contiguity.
- Minimize annexed areas.
- Minimize the number of Palestinian[s] affected.

---

[1] W. Clinton, 2000.

Security:

The key lies in an international presence that can only be withdrawn by mutual consent. This presence will also monitor the implementation of the agreement between both sides.

My best judgment is that the Israeli presence would remain in fixed locations in the Jordan Valley under the authority of the International force for another 36 months. This period could be reduced in the event of favorable regional developments that diminish the threats to Israel.

On early warning stations, Israel should maintain three facilities in the West Bank with a Palestinian liaison presence. The stations will be subject to review every 10 years with any changes in the status to be mutually agreed.

Regarding emergency developments, I understand that you will still have to develop a map of the relevant areas and routes. But in defining what is an emergency, I propose the following definition:

Imminent and demonstrable threat to Israel's national security of a military nature that requires the activation of a national state emergency.

Of course, the international forces will need to be notified of any such determination.

On airspace, I suggest that the state of Palestine will have sovereignty over its airspace but that two sides should work out special arrangements for Israeli training and operational needs.

I understand that the Israeli position is that Palestine should be defined as a "demilitarized state" while the Palestinian side proposes "a state with limited arms." As a compromise, I suggest calling it a "non-militarized state."

This will be consistent with the fact that in addition to a [sic] strong Palestinian security forces. [sic] Palestine will have an international force for border security and deterrent purposes.

Jerusalem and Refugees:

I have a sense that the remaining gaps have more to do with formulations than practical realities.

Jerusalem:

The general principle is that Arab areas are Palestinian and Jewish ones are Israeli. This would apply to the Old City as well. I urge the two sides to work on maps to create maximum contiguity for both sides.

Regarding the Haram/Temple Mount, I believe that the gaps are not related to practical administration but to the symbolic issues of sovereignty and to finding a way to accord respect to the religious beliefs of both sides.

I know you have been discussing a number of formulations, and you can agree [on] one of these. I add to these two additional formulations guaranteeing Palestinian effective control over the Haram while respecting the conviction of the Jewish people.

Regarding either one of these two formulations will be international monitoring to provide mutual confidence.

> 1—Palestinian sovereignty over the Haram, and Israeli sovereignty over a) the Western Wall and the space sacred to Judaism of which it is a part; b) the Western Wall and the Holy of Holies of which it is a part. There will be a fine commitment by both not to excavate beneath the Haram or behind the Wall.
> 2—Palestinian sovereignty over the Haram and Israeli sovereignty over the Western Wall and shared functional sovereignty over the issue of excavation under the Haram and behind the Wall such that mutual consent would be requested before any excavation can take place.

Refugees:

I sense that the differences are more relating to formulations and less to what will happen on a practical level.

I believe that Israel is prepared to acknowledge the moral and material suffering caused to the Palestinian people as a result of the 1948 war and the need to assist the international community in addressing the problem.

An international commission should be established to implement all the aspects that flow from your agreement: compensation, resettlement, rehabilitation, etc.

The US is prepared to lead an international effort to help the refugees.

The fundamental gap is on how to handle the concept of the right of return. I know the history of the issue and how hard it will be for the Palestinian leadership to appear to be abandoning this principle.

The Israeli side could not accept any reference to a right of return that would imply a right to immigrate to Israel in defiance of Israel's sovereign policies and admission or that would threaten the Jewish character of the state.

Any solution must address both needs.

The solution will have to be consistent with the two-state approach that both sides have accepted as a way to end the Palestinian-Israeli conflict: the state of Palestine as the homeland of the Palestinian people and the state of Israel as the homeland of the Jewish people.

Under the two-state solution, the guiding principle should be that the Palestinian state would be the focal point for Palestinians who choose to return to the area without ruling out that Israel will accept some of these refugees.

I believe that we need to adopt a formulation on the right of return that will make clear that there is no specific right of return to Israel itself but that does not negate the aspiration of the Palestinian people to return to the area.

In light of the above, I propose two alternatives:

> 1—Both sides recognize the right of Palestinian refugees to return to historic Palestine, or,
> 2—Both sides recognize the right of Palestinian refugees to return to their homeland.

The agreement will define the implementation of this general right in a way that is consistent with the two-state solution. It would list the five possible homes for the refugees:

> 1—The state of Palestine.
> 2—Areas in Israel being transferred to Palestine in the land swap.
> 3—Rehabilitation in host country.
> 4—Resettlement in third country.
> 5—Admission to Israel.

In listing these options, the agreement will make clear that the return to the West Bank, Gaza Strip, and areas acquired in the land swap would be the right of all Palestinian refugees, while rehabilitation in host countries, resettlement in third countries and absorption into Israel will depend upon the policies of those countries.

Israel could indicate in the agreement that it intends to establish a policy so that some of the refugees would be absorbed into Israel consistent with Israel's sovereign decision.

I believe that priority should be given to the refugee population in Lebanon.

The parties would agree that this implements resolution 194.

The End of Conflict:

I propose that the agreement clearly mark the end of the conflict and its implementation put an end to all claims. This could be implemented through a UN Security Counsel Resolution that notes that Resolutions 242 and 338 have been implemented and through the release of Palestinian prisoners.

Concluding remarks:

I believe that this is the outline of a fair and lasting agreement.

It gives the Palestinian people the ability to determine their future on their own land, a sovereign and viable state recognized by the international community, Al-Quds as its capital, sovereignty over the Haram, and new lives for the refugees.

It gives the people of Israel a genuine end to the conflict, real security, the preservation of sacred religious ties, the incorporation of 80% of the settlers into Israel, and the largest Jewish Jerusalem in history recognized by all as its capital.

Istanbul Cooperation Initiative[2]

1. With a transformed Alliance determined to respond to new challenges, NATO is ready to undertake a new initiative in the broader Middle East region to further contribute to long-term global and regional security and stability while complementing other international efforts.

2. In this context, progress towards a just, lasting, and comprehensive settlement of the Israeli-Palestinian conflict should remain a priority for the countries of the region and the international community as a whole, and for the success of the security and stability objectives of this initiative. Full and speedy implementation of the Quartet Road Map is a key element in international efforts to promote a two state solution to the Israeli-Palestinian conflict in which Israel and Palestine live side by side in peace and security. The roadmap is a vital element of international efforts to promote a comprehensive peace on all tracks, including the Syrian-Israeli and Lebanese-Israeli tracks.

[2] Istanbul Cooperation Initiative, 2004.

3. NATO's initiative, based on a series of mutually beneficial bilateral relationships aimed at fostering security and regional stability, should take into account the following principles:

a. the importance of taking into account ideas and proposals originating from the countries of the region or regional organisations;

b. the need to stress that the NATO initiative is a cooperative initiative, based on joint ownership and the mutual interests of NATO and the countries of the region, taking into account their diversity and specific needs;

c. the need to recognise that this process is distinct yet takes into account and complements other initiatives including by the G-8 and international organisations such as the EU and the OSCE as appropriate. The NATO initiative should also be complementary to the Alliance's Mediterranean Dialogue and could use instruments developed in this framework, while respecting its specificity. Furthermore, the new initiative could apply lessons learned and, as appropriate, mechanisms and tools derived from other NATO initiatives such as the Partnership for Peace (PfP);

d. the need to focus on practical cooperation in areas where NATO can add value, particularly in the security field. Participation of countries in the region in the initiative as well as the pace and extent of their cooperation with NATO will depend in large measure on their individual response and level of interest;

e. the need to avoid misunderstandings about the scope of the initiative, which is not meant to either lead to NATO/EAPC/PfP membership, provide security guarantees, or be used to create a political debate over issues more appropriately handled in other fora.

4. Taking into account other international efforts for reforms in the democracy and civil society fields in the countries of the region, NATO's offer to those countries of dialogue and cooperation will contribute to those efforts where it can have an added value: in particular, NATO could make a notable contribution in the security field as a result of its particular strengths and the experience gained with the PfP and the Mediterranean Dialogue.

Aim of the initiative

5. The aim of the initiative would be to enhance security and regional stability through a new transatlantic engagement with the region. This could be achieved by actively promoting NATO's cooperation with interested countries in the field of security, particularly through practical activities where NATO can add value to develop the ability of countries' forces to operate with those of the Alliance including by contributing to NATO-led operations, fight against terrorism, stem the flow

of WMD materials and illegal trafficking in arms, and improve countries' capabilities to address common challenges and threats with NATO.

6. Countries of the region [m]ight see benefit in cooperation with the Alliance through practical support against terrorist threats, access to training, defence reform expertise and opportunities for military cooperation, as well as through political dialogue on issues of common concern.

Content of the initiative including priority areas

7. The initiative's aim would be essentially achieved through practical cooperation and assistance in the following priority areas, and illustrative menu of specific activities:

a. providing tailored advice on defence reform, defence budgeting, defence planning and civil-military relations.

b. promoting military-to-military cooperation to contribute to interoperability[3] through participation in selected military exercises and related education and training activities that could improve the ability of participating countries' forces to operate with those of the Alliance in contributing to NATO-led operations consistent with the UN Charter:

- invite interested countries to observe and/or participate in selected NATO/PfP exercise activities as appropriate and provided that the necessary arrangements are in place;
- encourage additional participation by interested countries in NATO-led peace-support operations on a case-by-case basis;

c. fighting against terrorism including through information sharing and maritime cooperation:

- invite interested countries, in accordance with the procedures set out by the Council for contributory support from non-NATO nations, to join Operation Active Endeavour (OAE) in order to enhance the ability to help deter, defend, disrupt and protect against terrorism through maritime operations in the OAE Area of Operations;
- explore other forms of cooperation against terrorism including through intelligence exchange and assessments as appropriate.

[3] "Interoperability requirements constitute firm prerequisites for contributing nations such as the need to communicate with each other, to operate together, to support each other, and to train together" (footnote reprinted from the original document).

d. contributing to the work of the Alliance on threats posed by weapons of mass destruction (WMD) and their means of delivery:

e. promoting cooperation as appropriate and where NATO can add value in the field of border security, particularly in connection with terrorism, small arms & light weapons, and the fight against illegal trafficking:

- offer NATO-sponsored border security expertise and facilitate follow-up training in this respect;
- access to appropriate PfP programmes and training centres.

f. promoting cooperation in the areas of civil emergency planning:

- offer NATO training courses on civil emergency planning, civil-military coordination, and crisis response to maritime, aviation, and surface threats;
- invitations to join or observe relevant NATO/PfP exercises as appropriate and provision of information on possible disaster assistance.

Geographical scope of the initiative

8. Based on the principle of inclusiveness, the initiative could be opened to all interested countries in the region who subscribe to the aim and content of this initiative, including the fight against terrorism and the proliferation of weapons of mass destruction as described above. Each interested country would be considered by the North Atlantic Council on a case-by-case basis and on its own merit. This initiative would complement NATO's specific relationship with the partner countries of the Mediterranean Dialogue.[4]

Implementing the new initiative

9. This initiative would carry NATO into a new set of relationships with countries that may have a limited understanding of the Alliance as it has been transformed. Since an underlying requirement of success for the initiative is the development of ownership by countries of the region, it will be necessary to update governments' and opinion-formers' understanding of NATO and the initiative and, in the light of the reactions of the countries concerned, consider a joint public diplomacy effort. Furthermore, in developing and implementing the initiative, the views of interested countries in the region will have to be taken into account through a process of regular consultation.

[4] "Specificity in this respect refers in particular to the composition of this initiative and the Mediterranean Dialogue, as well as the multilateral dimension of the Mediterranean Dialogue" (footnote reprinted from the original document).

10. This initiative will be launched at the Istanbul Summit. Subsequently, in consultation with interested countries, NATO would offer a menu of practical activities within the above-mentioned priority areas for possible development with interested countries of the region. The Alliance would engage these countries, on a 26 + 1 basis, to develop and execute agreed work plans. While doing so, the new initiative could apply lessons learned and, as appropriate and on a case-by-case basis, mechanisms and tools derived from other NATO initiatives such as the Partnership for Peace (PfP). Appropriate legal, security and liaison arrangements should be put in place.

Treaty of Amity and Cooperation in Southeast Asia [Excerpt][5]

Chapter IV: Pacific Settlement of Disputes

Article 13

The High Contracting Parties shall have the determination and good faith to prevent disputes from arising. In case disputes on matters directly affecting them should arise, especially disputes likely to disturb regional peace and harmony, they shall refrain from the threat or use of force and shall at all times settle such disputes among themselves through friendly negotiations.

Article 14

To settle disputes through regional processes, the High Contracting Parties shall constitute, as a continuing body, a High Council comprising a Representative at ministerial level from each of the High Contracting Parties to take cognizance of the existence of disputes or situations likely to disturb regional peace and harmony.

Article 15

In the event no solution is reached through direct negotiations, the High Council shall take cognizance of the dispute or the situation and shall recommend to the parties in dispute appropriate means of settlement such as good offices, mediation, inquiry or conciliation. The High Council may however offer its good offices, or upon agreement of the parties in dispute, constitute itself into a committee of mediation, inquiry or conciliation. When deemed necessary, the High Council shall recommend appropriate measures for the prevention of a deterioration of the dispute or the situation.

---

5 Treaty of Amity and Cooperation in Southeast Asia, 1976.

Article 16

The foregoing provision of this Chapter shall not apply to a dispute unless all the parties to the dispute agree to their application to that dispute. However, this shall not preclude the other High Contracting Parties not party to the dispute from offering all possible assistance to settle the said dispute. Parties to the dispute should be well disposed towards such offers of assistance.

Article 17

Nothing in this Treaty shall preclude recourse to the modes of peaceful settlement contained in Article 33(l) of the Charter of the United Nations. The High Contracting Parties which are parties to a dispute should be encouraged to take initiatives to solve it by friendly negotiations before resorting to the other procedures provided for in the Charter of the United Nations.

Bibliography

Articles, Books, Web Pages, and Other Publications

"Afghan Reinforcements: Germany Pledges 500 Extra Troops Plus Big Aid Increase," Spiegel Online International, January 26, 2010. As of January 28, 2010:
http://www.spiegel.de/international/world/0,1518,674116,00.html

Afrasiabi, Kaveh L., "Iran Unveils a Persian Gulf Security Plan," Asia Times Online, April 14, 2007. As of November 2009:
http://www.atimes.com/atimes/Middle_East/ID14Ak04.html

Al-Juali, Mohamed Youssif, *Gulf Cooperation Council and Red Sea Security,* Dubai, United Arab Emirates: Gulf Research Center, April 2004.

Al-Khalifa, Khalid Bin Ahmed Bin Mohamed, "Statement by H. E. Shaikh Khalid Bin Ahmed Bin Mohamed Al-Khalifa Before the Sixty-Third Session of the United Nations General Assembly," New York, N.Y., September 27, 2008.

Al-Shayeji, Abdullah, "Dangerous Perceptions: Gulf Views of the U.S. Role in the Region," *Middle East Policy,* Vol. 5, 1997.

Armacost, Michael H., "U.S. Policy in the Persian Gulf and Kuwaiti Reflagging," statement before the Senate Foreign Relations Committee, Washington, D.C., June 16, 1987.

Armstrong, Glenda, *Terrorism 2007: Special Bibliography No. 332,* Maxwell Air Force Base, Ala.: Muir S. Fairchild Research Information Center, 2007.

ASEAN Secretariat, "Overview," Web page, undated. As of November 15, 2009:
http://www.aseansec.org/92.htm

———, "ASEAN Plus Three Cooperation," Web page, August 2009. As of November 15, 2009:
http://www.aseansec.org/16580.htm

Auswärtiges Amt, "Friends of Democratic Pakistan Group Meeting Set," Web page, October 31, 2008. As of November 15, 2009:
http://www.germany.info/Vertretung/usa/en/__PR/P__Wash/2008/10/31__Steinmeier__Gulf__PR.html

Ayson, Robert, "China Central? Australia's Asia Strategy," *International Spectator*, Vol. 44, No. 2, June 2009.

Ayub Khan, Mohammed, "The Pakistan-American Alliance," *Foreign Affairs*, January 1964.

Bakri, Nada, "Biden Warns of Ending Commitment," *Washington Post*, July 4, 2009.

Beehner, Lionel, "Russia-Iran Arms Trade," Council on Foreign Relations, Backgrounder, November 1, 2006. As of November 15, 2009:
http://www.cfr.org/publication/11869/#

Bensahel, Nora, "International Perspectives on Agency Reform: Testimony Presented Before the Armed Services Committee, Subcommittee on Oversight and Investigations on January 29, 2008," Santa Monica, Calif.: RAND Corporation, CT-298, 2008. As of November 9, 2009:
http://www.rand.org/pubs/testimonies/CT298/

Bensahel, Nora, Olga Oliker, and Heather Peterson, *Improving Capacity for Stabilization and Reconstruction Operations*, Santa Monica, Calif.: RAND Corporation, MG-852-OSD, 2009. As of November 9, 2009:
http://www.rand.org/pubs/monographs/MG852/

Berry, LaVerie, Glenn Curtis, and Rex Hudson, *Bibliography on Future Trends in Terrorism*, Washington, D.C.: Federal Research Division, Library of Congress, 1998.

Bloy, Marji, "George Canning (1770–1827)," The Victorian Web, March 5, 2002. As of November 15, 2009:
http://www.victorianweb.org/history/pms/canning.html

BP, *BP Statistical Review of World Energy*, June 2009.

Brinton, Crane, *The Anatomy of Revolution*, revised edition, New York, N.Y.: Vintage Books, 1965.

British American Security Information Council, "EU3 Negotiations with the Islamic Republic of Iran: Not Out of the Woods Yet and Time Is Short, Very Short," Basic Notes, July 2005. As of November 15, 2009:
http://www.basicint.org/pubs/Notes/BN050711.htm

Brzezinski, Zbigniew, *Power and Principle: Memoirs of the National Security Adviser, 1977–1981*, New York, N.Y.: Farrar, Strauss and Giroux, 1983.

Bumiller, Elisabeth, and Ellen Barry, "U.S. Searches for Alternative to Kyrgyz Base," *New York Times*, February 5, 2009.

Burke, Jason, "Revealed: Secret Taliban Peace Bid," *Observer* (London), September 28, 2009.

Bush, George H. W., "Toast at the State Dinner in Warsaw," Warsaw, Poland, July 10, 1989.

———, "Address Before a Joint Session of the Congress on the State of the Union," Washington, D.C., January 31, 1990.

Bush, George W., "Address Before a Joint Session of the Congress on the State of the Union," Washington, D.C., January 29, 2002a.

———, "Remarks by Mr. George W. Bush, President, at the International Conference on Financing for Development," Monterrey, Mexico, March 22, 2002b.

———, "Joint Understanding Read by President Bush at Annapolis Conference," Annapolis, Md., November 27, 2002c. As of November 15, 2009:
http://georgewbush-whitehouse.archives.gov/news/releases/2007/11/20071127.html

———, "President Discusses Roadmap for Peace in the Middle East: Remarks by the President on the Middle East," Washington, D.C.: The White House, March 14, 2003. As of November 15, 2009:
http://georgewbush-whitehouse.archives.gov/news/releases/2003/03/20030314-4.html

———, "Keynote Address," presented at the Saban Forum, "Strategic Choices: Challenges for the Next U.S. and Israeli Governments," Washington, D.C., December 5, 2008.

Bush, George W., Tony Blair, Jose Maria Aznar, and Jose Durao Barroso, press conference transcript, the Azores, Portugal, March 17, 2003. As of November 15, 2009:
http://www.guardian.co.uk/world/2003/mar/17/iraq.politics2

Carter, Jimmy, "The State of the Union Address Delivered Before a Joint Session of the Congress," Washington, D.C., January 23, 1980.

"China Floats Idea of First Overseas Naval Base," BBC News, December 30, 2009. As of January 20, 2010:
http://news.bbc.co.uk/2/hi/8435037.stm

Clinton, Hilary, "Remarks with Saudi Arabian Foreign Minister Prince Saud Al-Faisal," Washington, D.C., July 31, 2009.

Clinton, William J., "Statement on Signing the Iraq Liberation Act of 1998," The White House, Washington, D.C., October 31, 1998a.

———, "Statement by the President," The White House, Washington, D.C., December 16, 1998b. As of November 15, 2009:
http://clinton3.nara.gov/WH/New/html/19981216-3611.html

———, "Clinton Proposal on Israeli-Palestinian Peace," transcript of remarks at a White House meeting with U.S., Israeli, and Palestinian peace negotiators, Washington, D.C., December 23, 2000. Transcript supplied by the Jewish Peace Lobby, 2008. As of November 15, 2009:
http://www.peacelobby.org/clinton_parameters.htm

Clinton, William J., and Warren Christopher, "Clinton Issues Pledge to NPT Non-Nuclear Weapon States," Fact Sheet, April 6, 1995. As of January 20, 2010:
http://www.fas.org/nuke/control/npt/docs/940405-nsa.htm

"Cheney: No Link Between Saddam Hussein, 9/11," CNN.com, June 1, 2009. As of November 15, 2009:
http://www.cnn.com/2009/POLITICS/06/01/cheney.speech/

Cody, Edward, "First French Military Base Opens in the Persian Gulf," *Washington Post*, May 27, 2009.

Cohen, Herman J., "Somalia and the US Long and Troubled History," Web page, January 21, 2002. As of November 15, 2009:
http://www.raceandhistory.com/cgi-bin/forum/webbbs_config.pl/noframes/read/115

Cohen, William S., *Personal Accountability for Force Protection at Khobar Towers*, U.S. Department of Defense, July 31, 1997.

Cole, Juan, "Khamenei's Speech Replying to Obama," Web page, March 23, 2009. As of November 15, 2009:
http://www.juancole.com/2009/03/osc-khameneis-speech-replying-to-obama.html

"COMISAF Initial Assessment (Unclassified)—Searchable Document," *Washington Post*, September 21, 2009. As of November 15, 2009:
http://www.washingtonpost.com/wp-dyn/content/article/2009/09/21/AR2009092100110.html

Cordesman, Anthony H., and Khalid R. Al-Rodhan, *Gulf Military Forces in an Era of Asymmetric Warfare*, Westport, Conn.: Praeger, 2006.

Council of Arab States at the Summit Level, "The Arab Peace Initiative," official translation, Beirut, Lebanon, 2002. As of November 25, 2009:
http://www.al-bab.com/arab/docs/league/peace02.htm

Council of the European Union, "ESDP Structures and Instruments," Web page, undated-a. As of November 15, 2009:
http://www.consilium.europa.eu/showPage.aspx?id=279&lang=en

———, "EUMM Georgia," Web page, undated-b. As of November 15, 2009:
http://www.consilium.europa.eu/showPage.aspx?id=1512&lang=EN

Cragin, Kim, and Peter Chalk, *Terrorism and Development: Using Social and Economic Development to Inhibit a Resurgence of Terrorism*, Santa Monica, Calif.: RAND Corporation, MR-1630-RC, 2003. As of November 9, 2009:
http://www.rand.org/pubs/monograph_reports/MR1630/

CTBTO Preparatory Commission, home page, copyright 2008. As of November 15, 2009:
http://www.ctbto.org/

Curley, Tom, "Army Chief Says U.S. Ready to Be in Iraq 10 Years," Associated Press, May 26, 2009.

"Defense Agreement Signed Between Qatar and France," ArabicNews.com, October 26, 1998. As of November 15, 2009:
http://www.arabicnews.com/ansub/Daily/Day/981026/1998102612.html

"Defense Ties with Muslim States Underlined," *Iran Times,* April 11, 2007.

Dobbins, James, Sarah Harting, and Dalia Dassa Kaye, *Coping with Iran: Confrontation, Containment, or Engagement? A Conference Report*, Santa Monica, Calif.: RAND Corporation, CF-237-NSRD, 2007. As of November 9, 2009:
http://www.rand.org/pubs/conf_proceedings/CF237/

Dobbins, James, Seth G. Jones, Benjamin Runkle, and Siddarth Mohandas, *Occupying Iraq: A History of the Coalition Provisional Authority*, Santa Monica, Calif.: RAND Corporation, MG-847-CC, 2009. As of November 9, 2009:
http://www.rand.org/pubs/monographs/MG847/

Eggen, Dan, "9/11 Panel Links Al Qaeda, Iran: Bin Laden May Have Part in Khobar Towers, Report Says," *Washington Post*, June 26, 2004, p. A12.

El Kamel, Hussein, "Comments on the Future of the Mediterranean Dialogue," Palermo Atlantic Forum, Palermo, Italy, October 6, 2007.

Erlanger, Steven, and Katrin Bennhold, "Sarkozy Helps to Bring Syria Out of Isolation," *New York Times*, July 14, 2008.

EU Council Secretariat, "EU Naval Operation Against Piracy (EU NAVFOR Somalia—Operation ATALANTA)," March 2009. As of November 15, 2009:
http://ue.eu.int/uedocs/cmsUpload/090325FactsheetEUNAVFOR%20Somalia-version4_EN.pdf

European Union, *A Secure Europe in a Better World: European Security Strategy*, Brussels, Belgium, December 12, 2003.

Evans, Michael, "NATO Summit: The Truth Behind the Troops Heading to Afghanistan," Times Online, April 6, 2009. As of November 15, 2009:
http://www.timesonline.co.uk/tol/news/world/asia/article6040817.ece

"FACTBOX—The Strait of Hormuz, Iran and the Risk to Oil," Reuters, March 27, 2007.

Finley, Mark, "BP Statistical Review of World Energy," briefing, International Energy Forum Headquarters, Riyadh, Saudi Arabia, July 5, 2008.

Fischer, David Hackett, *Washington's Crossing (Pivotal Moments in American History)*, New York, N.Y.: Oxford University Press, 2005.

Flynn, Michael, Rich Juergens, and Thomas Cantrell, "Employing ISR: SOF Best Practices," *Joint Forces Quarterly*, No. 50, July 2008, pp. 56–61.

Foreign Ministers of the Gulf Cooperation Council Plus Two, "Full Text of GCC-Plus-Two Statement on Regional Security," printed in Iranfocus.com, January 17, 2007. As of November 15, 2009:
http://www.iranfocus.com/en/iran-general-/full-text-of-gcc-plus-two-statement-on-regional-security-09889.html

Forest, James J. F., *Terrorism and Counterterrorism: An Annotated Bibliography, Vol. 1*, West Point, N.Y.: Combating Terrorism Center, U.S. Military Academy, 2004.

Forest, James J. F., Thomas A. Bengston, Jr., Hilda Rosa Martinez, Nathan Gonzalez, and Bridget C. Nee, *Terrorism and Counterterrorism: An Annotated Bibliography, Vol. 2*, West Point, N.Y.: Combating Terrorism Center, U.S. Military Academy, 2006.

"France, Kuwait Sign Defense Agreements," Agence France Presse, December 4, 2006.

"France Opens Base in Abu Dhabi," United Press International, May 26, 2009.

Gannon, Kathy, "AP Interview: Karzai Willing to Talk to Taliban," Associated Press, December 3, 2009.

Garver, John, Flynt Leverett, and Hillary Mann Leverett, *Moving (Slightly) Closer to Iran: China's Shifting Calculus for Managing Its "Persian Gulf Dilemma*," Asia-Pacific Policy Papers Series, The Edwin O. Reischauer Center for East Asian Studies, 2009.

Gates, Robert, and James Cartwright, "DoD News Briefing with Secretary Gates and Gen. Cartwright from the Pentagon," transcript, Washington, D.C.: The Pentagon, September 17, 2009. As of November 15, 2009:
http://www.defenselink.mil/transcripts/transcript.aspx?transcriptid=4479

Gause, F. Gregory, III, *Relations between the Gulf Cooperation Council and the United States*, Dubai, United Arab Emirates: Gulf Research Center, January 2004.

———, "Saudi Arabia: Iraq, Iran and the Regional Power Balance and the Sectarian Question," *Strategic Insights*, February 2007a.

———, "Threats and Threat Perceptions in the Persian Gulf Region," *Middle East Policy*, Vol. XIV, No. 2, Summer 2007b.

General Secretariat of the Council of the EU, "Background: The High Representative for Foreign Affairs and Security Policy/The European External Action Service," press release, November 2009.

Glain, Stephan J., *Mullahs, Merchants, and Militants: The Collapse of Arab Democracy in the Arab World*, New York, N.Y.: Thomas Dunne Books/St. Martin's Press, 2004.

Glendinning, Lee, and James Sturcke, "Pirates Take Over Oil Tanker with British Crew on Board," guardian.co.uk, November 17, 2008. As of November 15, 2009:
http://www.guardian.co.uk/world/2008/nov/17/oil-tanker-pirates

GlobalSecurity.org, "Operation Northern Watch," Web page, April 27, 2005a. As of November 15, 2009:
http://www.globalsecurity.org/military/ops/northern_watch.htm

———, "Operation Southern Watch," Web page, April 27, 2005b. As of November 15, 2009:
http://www.globalsecurity.org/military/ops/southern_watch.htm

———, "Collective Security Treaty Organization (CSTO)," Web page, July 3, 2009c. As of January 15, 2009:
http://www.globalsecurity.org/military/world/int/csto.htm

Gompert, David C., John Gordon IV, Adam Grissom, David R. Frelinger, Seth G. Jones, Martin C. Libicki, Edward O'Connell, Brooke Stearns Lawson, and Robert E. Hunter, *War by Other Means—Building Complete and Balanced Capabilities for Counterinsurgency: RAND Counterinsurgency Study—Final Report*, Santa Monica, Calif.: RAND Corporation, MG-595/2-OSD, 2008a. As of November 9, 2009:
http://www.rand.org/pubs/monographs/MG595.2/

Gompert, David C., John Gordon IV, Adam Grissom, David R. Frelinger, Seth G. Jones, Martin C. Libicki, Edward O'Connell, Brooke Stearns Lawson, and Robert E. Hunter, *Countering Insurgency in the Muslim World: Rethinking U.S. Priorities and Capabilities*, Santa Monica, Calif.: RAND Corporation, RB-9326-OSD, 2008b. As of November 9, 2009:
http://www.rand.org/pubs/research_briefs/RB9326/

Gopal, Anand, "No Afghan-Taliban Peace Talks, for Now," *Christian Science Monitor*, October 9, 2008.

Gordon, Michael R., "War in the Gulf: Iraqi Air Force; Harboring of Iraqi Planes by Iran Calls Its Neutrality into Question," *New York Times*, January 29, 1991.

Graham, Thomas, Jr., and Keith A. Hansen, *Spy Satellites and Other Intelligence Technologies That Changed History*, Seattle, Wash.: University of Washington Press, 2007.

Green, Jerrold D., Frederic Wehrey, and Charles Wolf, Jr., *Understanding Iran*, Santa Monica, Calif.: RAND Corporation, MG-771-SRF, 2009. As of November 9, 2009:
http://www.rand.org/pubs/monographs/MG771/

Green, Matthew, "FT Interview Transcript: Gen Stanley McChrystal," FT.com, January 25, 2010. As of January 25, 2010:
http://www.ft.com/cms/s/0/1036aae6-074a-11df-a9b7-00144feabdc0.html

Gromov, Boris, and Dmitry Rogozin, "Russian Advice on Afghanistan," *New York Times*, January 11, 2010.

"Gulf States Launch Arab Aid Plan to Rebuild Gaza," Reuters, February 22, 2009.

"Gulf States Urge Peace with Iran," BBC News, December 4, 2007. As of November 15, 2009:
http://news.bbc.co.uk/2/hi/middle_east/7127451.stm

Gwertzman, Bernard, "Hunter: Russia Is Long Run 'Loser' in Georgia Conflict," Council on Foreign Relations, September 3, 2008.

———, "Obama May Face 'Rebuff' from Europe on Military Step-Up in Afghanistan," Council on Foreign Relations, January 22, 2009.

"Hamas Coup in Gaza: Fundamental Shift in Palestinian Politics," *IIS Strategic Comments*, Vol. 13, No. 5, June 2007.

Hamzeh, Alia Shukri, "Jordan Offers to Host NATO-Supported Regional Security Cooperation Centre," *Jordan Times*, January 7, 2007.

Harahan, Joseph P., and John C. Kuhn III, *On-Site Inspections Under the CFE Treaty: A History of the On-Site Inspection Agency and CFE Treaty Implementation, 1990–1996*, Washington, D.C.: The On-Site Inspection Agency, U.S. Department of Defense, 1996.

"Hostage Captain Rescue; Navy Snipers Kill 3 Pirates," CNN, April 12, 2009.

Howard, Michael, and Robert Hunter, *Israel and the Arab World: The Crisis of 1967*, Adelphi Paper No. 41, Institute for Strategic Studies, September 1967.

———, "The Soviet Dilemma in the Middle East, Part I: Problems of Commitment," Adelphi Paper No. 59, Institute for Strategic Studies, 1969a.

———, "The Soviet Dilemma in the Middle East, Part II: Oil and the Persian Gulf," Adelphi Paper No. 60, Institute for Strategic Studies, 1969b.

Hunter, Robert E., *The European Security and Defense Policy: NATO's Companion—or Competitor?* Santa Monica, Calif.: RAND Corporation, MR-1463-NDRI/RE, 2002. As of November 9, 2009: http://www.rand.org/pubs/monograph_reports/MR1463/

———, "A Forward Looking Partnership," *Foreign Affairs*, September/October 2004.

———, "A 'Europe Whole and Free and at Peace,'" *Providence Journal*, September 9, 2008a.

———, "A New Grand Strategy for the United States: Testimony Presented Before the House Armed Services Committee, Subcommittee on Oversight and Investigations on July 31, 2008," Santa Monica, Calif.: RAND Corporation, CT-313, 2008b. As of November 9, 2009: http://www.rand.org/pubs/testimonies/CT313/

———, "A New American Middle East Strategy?" *Survival*, Vol. 50, No. 6, December 2008–January 2009, pp. 49–66.

———, "NATO After the Summit: Rebuilding Consensus: Testimony Presented Before the Senate Foreign Relations Committee, Subcommittee on European Affairs on May 6, 2009," Santa Monica, Calif.: RAND Corporation, CT-331, 2009. As of November 9, 2009: http://www.rand.org/pubs/testimonies/CT331/

Hunter, Robert E., Edward Gnehm, and George Joulwan, *Integrating Instruments of Power and Influence: Lessons Learned and Best Practices*, Santa Monica, Calif.: RAND Corporation, CF-251-NDF/KAF/RF/SRF, 2008. As of November 9, 2009: http://www.rand.org/pubs/conf_proceedings/CF251/

Hunter, Robert E., and Seth G. Jones, *Building a Successful Palestinian State: Security*, Santa Monica, Calif.: RAND Corporation, MG-146/2-DCR, 2006. As of November 9, 2009: http://www.rand.org/pubs/monographs/MG146.2/

Hunter, Robert E., and Khalid Nadiri, *Integrating Instruments of Power and Influence in National Security: Starting the Dialogue*, Santa Monica, Calif.: RAND Corporation, CF-231-CC, 2006. As of November 9, 2009: http://www.rand.org/pubs/conf_proceedings/CF231/

Hunter, Shireen T., ed., *Islam, Europe's Second Religion: The New Social, Cultural, and Political Landscape*, Westport, Conn.: Praeger, 2002.

———, ed., *Reformist Voices of Islam: Mediating Islam and Modernity*, Armonk, N.Y.: M. E. Sharpe, 2009.

Ibrahim, Alia, "French President, on Visit to Syria, Calls for Improved Ties Between Nations," *Washington Post*, September 4, 2008, p. A6.

"India, U.S. Agree on End User Monitoring Pact," *Times of India*, July 20, 2009.

"Indian Warships in Persian Gulf," *Times of India* (online), September 12, 2004. As of November 15, 2009: http://timesofindia.indiatimes.com/articleshow/848249.cms

International Atomic Energy Agency, home page, copyright 2003–2009. As of November 15, 2009: http://www.iaea.org

International Institute for Strategic Studies, *The Military Balance 2009*, UK: Routledge, 2009.

International Security Assistance Force—Afghanistan, "Troop Contributing Nations," (last updated December 22, 2009). As of January 15, 2010:
http://www.isaf.nato.int/en/troop-contributing-nations-3.html

"Interview of Russian Minister of Foreign Affairs Sergey Lavrov," *Al-Hayat*, August 2008.

Islamic Republic of Iran, "Cooperation for Peace, Justice and Progress: Package of Proposals by the Islamic Republic of Iran for Comprehensive and Constructive Negotiations," September 9, 2009.

Isseroff, Ami, "Quartet Roadmap to Israeli-Palestinian Peace: Introduction," MidEastWeb, April 30, 2003. As of November 15, 2009:
http://www.mideastweb.org/quartetrm3.htm

Ito, Shingo, "New Japan PM Offers Obama Help on Afghanistan," Agence France Presse, September 24, 2009.

Jajacobs, "Documenting the Government—Strait of Hormuz Edition," blog post on Free Government Information, January 11, 2008. As of November 15, 2009:
http://freegovinfo.info/node/1567

"Japan to Dispatch Reconstruction Team to Afghan," Press Trust of India, January 9, 2009.

Jones, Seth G., *Counterinsurgency in Afghanistan: RAND Counterinsurgency Study, Vol. 4*, Santa Monica, Calif.: RAND Corporation, MG-595-OSD, 2008. As of November 9, 2009:
http://www.rand.org/pubs/monographs/MG595/

Kaim, Markus, ed., *Great Powers and Regional Order: The United States in the Persian Gulf*, Hampshire, England, and Burlington, Vt.: Ashgate Publishing Limited, 2008.

Kaye, Dalia Dassa, *Talking to the Enemy: Track Two Diplomacy in the Middle East and South Asia*, Santa Monica, Calif.: RAND Corporation, MG-592-NSRD, 2007. As of November 9, 2009:
http://www.rand.org/pubs/monographs/MG592/

Kaye, Dalia Dassa, and Frederic M. Wehrey, "A Nuclear Iran: The Reactions of Neighbours, *Survival*, Vol. 49, No. 2, June 2007, pp. 111–128.

Kaye, Dalia Dassa, Frederic Wehrey, Audra K. Grant, and Dale Stahl, *More Freedom, Less Terror? Liberalization and Political Violence in the Arab World*, Santa Monica, Calif.: RAND Corporation, MG-772-RC, 2008. As of November 9, 2009:
http://www.rand.org/pubs/monographs/MG772/

Keinon, Herb, and Associated Press, "'Russia to Rethink S-300 Sale to Iran,'" *Jerusalem Post* (online), August 19, 2009. As of November 15, 2009:
http://www.jpost.com/servlet/Satellite?pagename=JPost/JPArticle/ShowFull&cid=1249418644694

Kelly, Terrence K., "An Iraqi Modus Vivendi: How Would It Come About and What Would It Look Like? Testimony Presented Before the Senate Foreign Relations Committee on April 3, 2008," Santa Monica, Calif.: RAND Corporation, CT-303, April 3, 2008. As of November 9, 2009:
http://www.rand.org/pubs/testimonies/CT303/

Kennedy, Edward M., "The Persian Gulf: Arms Race or Arms Control," *Foreign Affairs*, October 1975.

Kershner, Isabel, "Fatah Party Election Brings in a New Generation, *New York Times,* August 11, 2009.

Kessler, Glenn, "U.S., India Set Up 'Strategic Dialogue,'" *Washington Post*, July 21, 2009.

Khan, Mohammed Ayub, "The Pakistan-American Alliance," *Foreign Affairs*, January 1964.

Kinnunen, Mikko, "The European Security and Defence Policy (ESDP); Finland's EU Presidency; The Case of Israeli-Lebanon Conflict and the ESDP," briefing, Bratislava, Slovakia, June 2007.

Knowlton, Brian, "Biden Suggests U.S. Not Standing in Israel's Way on Iran," *New York Times*, July 5, 2009.

Kurata, Phillip, "Former U.S. Envoy to Afghanistan Reviews Bonn Agreement Success," Washington File, October 6, 2005.

Lamb, Christina, "Taliban Chief Backs Afghan Peace Talks," *Sunday Times* (London), March 15, 2009.

Lefebvre, Jeffrey Alan, *Arms for the Horn: U.S. Security Policy in Ethiopia and Somalia, 1953–1991*, Pittsburgh, Pa.: University of Pittsburgh Press, 1991.

Legrenzi, Matteo, "NATO in the Gulf: Who Is Doing Whom a Favor? *Middle East Policy*, Spring 2007.

Luciani, Giacomo, and Felix Neugart, *The European Union and the Gulf Cooperation Council: Towards a New Partnership*, Dubai, United Arab Emirates: Gulf Research Center, 2006.

Machefsky, Ira, "Piracy Threatens Oil Transit through Gulf of Aden," TheNumbersGuru.com, September 10, 2008. As of November 15, 2009:
http://thenumbersguru.blogspot.com/2008/09/piracy-threatens-oil-transit-through.html

Mackay, Neil, "Pirates of the Persian Gulf," *Sunday Herald* [Glasgow], October 27, 2007.

Mattair, Thomas R., "Mutual Threat Perceptions in the Arab/Persian Gulf: GCC Perceptions in the Arab/Persian Gulf: GCC Perceptions," *Middle East Policy Council Journal*, Vol. XIV, No. 2, Summer 2007.

McMillan, Richard Sokolsky, and Andrew C. Winner, "Toward a New Regional Security Architecture," *Washington Quarterly*, Vol. 26, No. 3, Summer 2003, pp. 161–175.

Miller, Rory, and Ashraf Mishrif, "The Barcelona Process and Euro-Arab Economic Relations: 1995–2005," *Middle East Review of International Affairs*, Vol. 9, No. 2, June 2005.

Ministère de la Défense, "General Stéphane Abrial, Chief of Staff of the French Air Force," October 2009. As of November 15, 2009:
http://www.defense.gouv.fr/air_uk/content/download/42536/425646/file/biographie_us_du_cemaa_bio_gaa_abrial_us_octo_2006.pdf

Ministry for Culture and Heritage, "Nuclear-Free Legislation—Nuclear-Free New Zealand," Web page, August 5, 2008. As of November 15, 2009:
http://www.nzhistory.net.nz/politics/nuclear-free-new-zealand/nuclear-free-zone

Ministry of Foreign Affairs of Japan, "Japan-Afghanistan Relations," Web page, November 2009. As of November 15, 2009:
http://www.mofa.go.jp/region/middle_e/afghanistan/

Missiroli, Antonio, *The Impact of the Lisbon Treaty on ESDP*, Brussels, Belgium: European Communities, 2008.

Moore, Molly, "France Announces Base in Persian Gulf," *Washington Post*, Wednesday, January 16, 2008, p. A11.

Morley, Jefferson, "Israeli Withdrawal from Gaza Explained," *Washington Post*, August 10, 2005.

Moss, Trefor, and Farhan Bokhari, "US Military Opens Transit Route Between Russia and Afghanistan," *Jane's Defence Weekly*, March 6, 2009.

Motes, Kevin D., *Annotated Bibliography of Government Documents Related to the Threat of Terrorism and the Attacks of September 11, 2001*, Oklahoma City, Okla.: U.S. Government Information Division, Oklahoma Department of Libraries, 2004.

Mozgovaya, Natasha, "Top Obama Aide: U.S. Commitment to Israel Is Not a Slogan," Haaretz.com, October 27, 2009. As of November 15, 2009:
http://www.haaretz.com/hasen/spages/1123992.html

Netanyahu, Benjamin, "Address by Prime Minister Benjamin Netanyahu at the Conference of the Organization on Security and Cooperation in Europe (OSCE)," Lisbon, Portugal, December 3, 1996.

———, "Full Text of Netanyahu's Foreign Policy Speech at Bar Ilan," printed in Haaretz.com, June 14, 2009. As of November 15, 2009:
http://www.haaretz.com/hasen/spages/1092810.html

"No More Gestures to Saudis: Iraq's Maliki," Agence France Presse, May 28, 2009.

North Atlantic Treaty Organization, "Bio Deputy Commander KFOR," Web page, undated-a (last updated August 11, 2008). As of November 15, 2009:
http://www.nato.int/kfor/structur/whoswho/cv/bio_stoltz_new.htm

North Atlantic Treaty Organization, "Civil Emergency Planning," Web page, undated-b (last updated October 23, 2009). As of November 15, 2009:
http://www.nato.int/cps/en/natolive/topics_49158.htm

North Atlantic Treaty Organization, "Education and Training," Web page, undated-c (last updated July 5, 2007). As of November 15, 2009:
http://www.nato.int/issues/education_and_training/participates.html

North Atlantic Treaty Organization, "The Euro-Atlantic Partnership Council," Web page, undated-d (last updated October 23, 2009). As of November 15, 2009:
http://www.nato.int/cps/en/natolive/topics_49276.htm

North Atlantic Treaty Organization, "ISAF Troop Contributing Nations," Web page, undated-e (last updated October 21, 2009). As of November 15, 2009:
http://www.nato.int/isaf/structure/nations/index.html

North Atlantic Treaty Organization, "NATO Training Mission–Iraq: Participating Nations," Web page, undated-f. As of November 15, 2009:
http://www.afsouth.nato.int/JFCN_Missions/NTM-I/Factsheets/NTMI_part.htm

North Atlantic Treaty Organization, "Senior Civil Emergency Planning Committee (SCEPC)," Web page, undated-g (last updated September 19, 2007). As of November 15, 2009:
http://www.nato.int/issues/scepc/index.html

North Atlantic Treaty Organization, "NATO Elevates Mediterranean Dialogue to a Genuine Partnership, Launches Istanbul Cooperation Initiative," Web page, June 29, 2004 (last updated October 16, 2008). As of November 15, 2009:
http://www.nato.int/docu/update/2004/06-june/e0629d.htm

North Atlantic Treaty Organization, "NATO Launches Training Initiative for Mediterranean and Middle East," Web page, November 29, 2006 (last updated November 30, 2006). As of November 15, 2009:
http://www.nato.int/docu/update/2006/11-november/e1129f.htm

North Atlantic Treaty Organization, "NATO Resumes Counter-Piracy Mission," Web page, March 24, 2009a (last updated November 10, 2009). As of November 15, 2009:
http://www.nato.int/cps/en/natolive/news_52016.htm?selectedLocale=en

North Atlantic Treaty Organization, "NATO Training Mission—Afghanistan," Web page, April 4, 2009b (last updated August 18, 2009). As of November 15, 2009:
http://www.nato.int/cps/en/natolive/news_52802.htm

Obama, Barack, "Responsibly Ending the War in Iraq," remarks, Camp Lejeune, N.C., February 27, 2009a. As of November 15, 2009:
http://www.whitehouse.gov/the_press_office/
Remarks-of-President-Barack-Obama-Responsibly-Ending-the-War-in-Iraq/

———, "Videotaped Remarks by the President in Celebration of Nowruz," transcript, Washington, D.C.: The White House, Office of the Press Secretary, March 20, 2009b. As of November 15, 2009:
http://www.whitehouse.gov/the_press_office/
videotaped-remarks-by-the-president-in-celebration-of-nowruz/

———, "Remarks by the President on a New Beginning," Cairo University, Cairo, Egypt, June 4, 2009c. As of November 15, 2009:
http://www.whitehouse.gov/the_press_office/Remarks-by-the-President-at-Cairo-University-6-04-09/

———, "Press Conference by the President," remarks, L'Aquila, Italy: U.S. Press Filing Center, July 10, 2009d. As of November 15, 2009:
http://www.whitehouse.gov/the_press_office/
Press-Conference-by-the-President-in-LAquila-Italy-7-10-09/

———, "Remarks by the President in Address to the Nation on the Way Forward in Afghanistan and Pakistan," West Point, New York, December 1, 2009e.

Obama, Barack, and Yukio Hatoyama, "Remarks by President Obama and Prime Minister Hatoyama of Japan After Bilateral Meeting," New York, N.Y., September 23, 2009. As of November 15, 2009:
http://www.whitehouse.gov/the_press_office/
Remarks-By-President-Obama-And-Prime-Minister-Hatoyama-of-Japan-After-Bilateral-Meeting/

Obama, Barack, and Benjamin Netanyahu, "Remarks by President Obama and Prime Minister Netanyahu of Israel in Press Availability," Washington, D.C.: The White House, May 18, 2009. As of November 15, 2009:
http://www.whitehouse.gov/the_press_office/
Remarks-by-President-Obama-and-Israeli-Prime-Minister-Netanyahu-in-press-availability/

O'Hanlon, Michael, "Estimating Casualties in a War to Overthrow Saddam," *Orbis,* Vol. 47, No. 1, Winter 2003, pp. 21–40.

Organisation of the Islamic Conference, "Member States," Web page, undated. As of November 15, 2009:
http://www.oic-oci.org/member_states.asp

Organization of the Petroleum Exporting Countries, "Who Are OPEC Member Countries," Web page, undated. As of November 15, 2009:
http://www.opec.org/library/faqs/aboutopec/q3.htm

"OSCE Working to Assist Afghanistan," UPI, September 10, 2008.

Palestine Facts, "What Was the Khartoum Resolution of 1967?" Web page, undated. As of November 15, 2009:
http://www.palestinefacts.org/pf_1967to1991_khartoum.php

Palestinian Peace Coalition and Geneva Initiative, "The Geneva Accord: A Model Israeli-Palestinian Peace Agreement," Web page, undated. As of November 15, 2009:
http://www.geneva-accord.org/mainmenu/english

Palmowski, Jan, "Mutual and Balanced Force Reduction Talks," printed in Encyclopedia.com, undated, originally published in *A Dictionary of Contemporary World History*, Oxford University Press, 2004. As of November 15, 2009:
http://www.encyclopedia.com/doc/1O46-MutualndBlncdFrcRdctnTlks.html

Pan, Esther, "Iraq: Madrid Donor Conference," Council on Foreign Relations, Backgrounder, October 30, 2003. As of November 15, 2009:
http://www.cfr.org/publication/7682/

Perry, Walter L., Stuart E. Johnson, Keith Crane, David C. Gompert, John Gordon IV, Robert E. Hunter, Dalia Dassa Kaye, Terrence K. Kelly, Eric Peltz, and Howard J. Shatz, *Withdrawing from Iraq: Alternative Schedules, Associated Risks, and Mitigating Strategies*, Santa Monica, Calif.: RAND Corporation, MG-882-OSD, 2009. As of November 9, 2009:
http://www.rand.org/pubs/monographs/MG882/index.html

Pflanz, Mike, "*Sirius Star* Oil Tanker Released After £2m Ransom Paid," Telegraph.co.uk, January 9, 2009. As of November 15, 2009:
http://www.telegraph.co.uk/news/worldnews/piracy/4208438/Sirius-Star-oil-tanker-released-after-2m-ransom-paid.html

"PGCC to Talk with Iran on Establishment of Common Market," Islamic Republic News Agency, April 27, 2008.

"Proposals for Persian Gulf Security: Build Trust, Cooperation," *Iran Times*, April 11, 2007.

RAND Palestinian State Study Team, *Building a Successful Palestinian State*, Santa Monica, Calif.: RAND Corporation, MG-146-1-DCR, 2007a. As of November 9, 2009:
http://www.rand.org/pubs/monographs/MG146-1/

———, *Helping a Palestinian State Succeed: Key Findings*, Santa Monica, Calif.: RAND Corporation, MG-146/1-1-RC, 2007b. As of November 9, 2009:
http://www.rand.org/pubs/monographs/MG146.1-1/

Rasmussen, Anders Fogh, "First NATO Press Conference," Brussels, Belgium, August 3, 2009a.

———, "Press Conference," December 4, 2009b. As of December 21, 2009:
http://www.nato.int/cps/en/natolive/opinions_59872.htm

———, "Statement by NATO Secretary General on Force Generation for Afghanistan," December 7, 2009c.

Rathmell, Andrew, Theodore W. Karasik, and David C. Gompert, *A New Persian Gulf Security System*, Santa Monica, Calif.: RAND Corporation, IP-248-CMEPP, 2003. As of November 9, 2009:
http://www.rand.org/pubs/issue_papers/IP248/

Reid, Tim, "Iraq Seeks to Buy U.S.-Made F-16 Fighter Jets," *The Times* (London), April 2, 2009.

Rempel, John, "A Brief History of United Nations Sanctions Against Iraq," The Peace and Justice Support Network of Mennonite Church USA, undated. As of November 15, 2009:
http://peace.mennolink.org/articles/iraqsancthist.html

Reynolds, Camille, *Features—Bibliography on Terrorism, Bioterrorism, the Middle East, and 9-11 Related Issues*, San Francisco, Calif.: Nossaman Guthner Knox & Elliott, 2001.

Reynolds, Paul, "New Russian World Order: The Five Principles," BBC News, September 1, 2008. As of November 15, 2009:
http://news.bbc.co.uk/2/hi/europe/7591610.stm

Rickover, Hyman G., *How the Battleship Maine Was Destroyed*, Washington, D.C.: Department of the Navy, Naval History Division, 1976.

Robertson, David, "Dubai Plans $200bn Canal to Bypass Strait of Hormuz," *The Times* (London), September 9, 2008.

Rosenberg, Jennifer, "Top 5 Crimes of Saddam Hussein," About.com, undated. As of November 15, 2009:
http://history1900s.about.com/od/saddamhussein/a/husseincrimes.htm

Rowhani, Hassan, "10-Point Plan to Promote 'Cooperation, Security, and Development' in Persian Gulf," World Economic Forum, Doha, Qatar, April 9–10, 2007, printed in Center for Strategic Research, April 14, 2007. As of November 15, 2009:
http://www.csr.ir/departments.aspx?lng=en&abtid=00&depid=106&semid=193

The Royal Forums, "Qaboos' 3 Day Visit to Tehran," Web page, August 8, 2009. As of November 15, 2009:
http://www.theroyalforums.com/10477-qaboos-3-day-visit-to-tehran/

Rumsfeld, Donald H., press conference, transcript, Brussels, Belgium, June 6, 2002. As of November 15, 2009:
http://www.defenselink.mil/transcripts/transcript.aspx?transcriptid=3490

Russell, James A., "Mutual Threat Perceptions in the Arab/Persian Gulf: GCC Perceptions," *Middle East Policy*, Vol. XIV, No. 2, Summer 2007.

Russell, Richard L., "The Persian Gulf's Collective-Security Mirage," *Middle East Policy*, Vol. 12, No. 2, Winter, 2005, pp. 77–88.

"Russia Approves U.S. Military Overflights to Afghanistan," VOANews.com, July 6, 2009. As of November 15, 2009:
http://www.voanews.com/english/2009-07-06-voa35.cfm

Sadikhova, U., "Relations with Israel Will Be Re-Established Only After Adoption of Arab Peace Initiative: OIC Head," *The Journal of Turkish Weekly*, May 25, 2009.

Sanger, David, and Eric Schmitt, "U.S. Speeding Up Missile Defenses in Persian Gulf," *New York Times*, January 30, 2010.

Sarkozy, Nicolas, "Fifteenth Ambassadors' Conference: Speech by M. Nicolas Sarkozy, President of the Republic," Paris, France, August 27, 2007.

———, "Discours lors du Colloque de la Fondation pour la Recherche Stratégique «La France, la Défense Européenne et l'OTAN au XXIème»," Paris, France, March 11, 2009.

Scheffer, Jaap de Hoop, "Speech," Israel, February 24, 2005.

The Shanghai Cooperation Organisation, homepage, undated. As of January 8, 2010:
http://www.sectsco.org/EN/#

Sick, Gary, *All Fall Down: America's Tragic Encounter with Iran*, New York, N.Y.: Random House, 1986.

———, "Twin Pillars to Dual Containment," in David W. Lesch, ed., *The Middle East and the United States: A Historical and Political Reassessment*, Second Edition, Boulder, Colo., and Oxford, UK: Westview Press, 2003.

Solana, Javier, "ESDP@10: What Lessons for the Future," Council of the European Union, S195/09, July 28, 2009.

"Spain and Britain Agree on Closer Cooperation After Historic Gibraltar Visit," *Deutsche Welle*, July 21, 2009.

"Sultan Qaboos to Visit Iran," *Khaleej Times* (Dubai), June 18, 2009.

Tatchell, Peter, "Ahmadinejad Accepts Israel's Right to Exist," guardian.co.uk, September 29, 2008. As of November 15, 2009:
http://www.guardian.co.uk/commentisfree/2008/sep/29/iran.israel.ahmadinejad

"Text of Final Communique [sic] of the 28th GCC Summit," printed in GulfNews.com, December 5, 2007. As of November 15, 2009:
http://gulfnews.com/news/gulf/uae/general/
text-of-final-communique-of-the-28th-gcc-summit-1.217048

Thaler, David E., Theodore W. Karasik, Dalia Dassa Kaye, Jennifer D. P. Moroney, Frederic Wehrey, Obaid Younossi, Farhana Ali, and Robert A. Guffey, *Future U.S. Security Relationships with Iraq and Afghanistan: U.S. Air Force Roles*, Santa Monica, Calif.: RAND Corporation, MG-681-AF, 2007. As of November 9, 2009:
http://www.rand.org/pubs/monographs/MG681/

Tyler, Patrick E., "U.S. Strategy Plan Calls for Insuring No Rivals Develop," *New York Times*, March 8, 1992.

United Kingdom Foreign & Commonwealth Office, "Outcomes from Afghanistan: The London Conference," Web page, 2010. As of January 28, 2010:
http://afghanistan.hmg.gov.uk/en/conference/outcomes/

United Nations Development Programme, "Afghanistan's New Beginnings Programme (ANBP)," Web page, June 2009. As of December 2009:
http://www.undp.org.af/WhoWeAre/UNDPinAfghanistan/Projects/psl/prj_anbp.htm

The United States and the Soviet Union, "Letter of Invitation to Madrid Peace Conference," October 30, 1991.

United States Institute of Peace, "Iraqi Prime Minister Will Keep Door Open for U.S. Military Role After 2011," Web page, July 23, 2009. As of November 15, 2009:
http://www.usip.org/newsroom/news/iraqi-prime-minister-will-keep-door-open-us-military-role-after-2011

U.S. Department of State, "The Baghdad Pact (1955) and the Central Treaty Organization (CENTO)," Web page, undated. As of November 15, 2009:
http://www.state.gov/r/pa/ho/time/lw/98683.htm.

———, "Background Note: Oman," Web page, June 2007. As of November 15, 2009:
http://www.state.gov/r/pa/ei/bgn/35834.htm

U.S. Department of State, Office of the Spokesman, "A Performance-Based Roadmap to a Permanent Two-State Solution to the Israeli-Palestinian Conflict," Washington, D.C., April 30, 2003.

"U.S. to Join Nuclear Talks with Iran, State Department Says," CNN, April 8, 2009.

"U.S. 'Will Repel Nuclear Hopefuls,'" BBC News, July 22, 2009. As of November 15, 2009:
http://news.bbc.co.uk/2/hi/asia-pacific/8162402.stm

Vlahos, Michael, "Fighting Identity: Why We Are Losing Our Wars," *Military Review*, November–December 2007, pp. 2–12.

Warrick, Joby, "U.S. Steps Up Arms Sales to Persian Gulf Allies," *Washington Post*, January 31, 2010.

Wehrey, Frederic, Theodore W. Karasik, Alireza Nader, Jeremy Ghez, Lydia Hansell, and Robert A. Guffey, *Saudi-Iranian Relations Since the Fall of Saddam: Rivalry, Cooperation, and Implications for U.S. Policy*, Santa Monica, Calif.: RAND Corporation, MG-840-SRF, 2009. As of November 9, 2009:
http://www.rand.org/pubs/monographs/MG840/

Wehrey, Frederic, David E. Thaler, Nora Bensahel, Kim Cragin, Jerrold D. Green, Dalia Dassa Kaye, Nadia Oweidat, and Jennifer Li, *Dangerous But Not Omnipotent: Exploring the Reach and Limitations of Iranian Power in the Middle East*, Santa Monica, Calif.: RAND Corporation, MG-781-AF, 2009a. As of November 9, 2009:
http://www.rand.org/pubs/monographs/MG781/

———, *A New U.S. Policy Paradigm Toward Iran*, Santa Monica, Calif.: RAND Corporation, RB-9422-AF, 2009b. As of November 9, 2009:
http://www.rand.org/pubs/research_briefs/RB9422/

Wheeler, Julia, "Gulf States Form Common Market," BBC News, January 1, 2008. As of November 15, 2009:
http://news.bbc.co.uk/2/hi/middle_east/7166549.stm

The White House, Office of the Press Secretary, "What's New in the Strategy for Afghanistan and Pakistan," news release, Washington, D.C., March 27, 2009. As of November 15, 2009:
http://www.whitehouse.gov/the_press_office/
Whats-New-in-the-Strategy-for-Afghanistan-and-Pakistan/

Wilson, Scott, "Hamas Sweeps Palestinian Elections, Complicating Peace Efforts in Mideast," *Washington Post*, January 27, 2006.

Yaffe, Michael, "An Overview of the Middle East Peace Process: Working Group on Arms Control and Regional Security," in F. Tanner, ed., *Arms Control, Confidence-Building and Security Cooperation in the Mediterranean, North Africa, and the Middle East*, Malta: The Mediterranean Academy of Diplomatic Studies, 2004a.

———, "The Gulf and a New Middle East Security System," *Middle East Policy*, Vol. 11, No. 3, Fall 2004b, pp. 85–91.

Charters, Resolutions, Treaties, and Related Documents

Agreement Between the Government of the United States of America and the Government of the Union of Soviet Socialist Republics on the Prevention of Incidents on and over the High Seas, Moscow, Soviet Union, May 25, 1972.

Agreement Between the United States of America and the Republic of Iraq on the Withdrawal of United States Forces from Iraq and the Organization of Their Activities During Their Temporary Presence in Iraq, Baghdad, Iraq, November 17, 2008.

Agreement on Provisional Arrangements in Afghanistan Pending the Re-Establishment of Permanent Government Institutions, participants in the UN Talks on Afghanistan, Bonn, Germany, December 5, 2001.

ASEAN Declaration, Bangkok, Thailand, August 8, 1967.

Barcelona Declaration, adopted at the Euro-Mediterranean Conference, November 27–28, 1995.

Basic Document of the Euro-Atlantic Partnership Council, Sintra, Portugal, May 30, 1997. As of December 2, 2009:
http://www.nato.int/cps/en/SID-4124E21F-BB820EC6/natolive/official_texts_25471.htm

Bucharest Summit Declaration, North Atlantic Treaty Organization, April 3, 2008.

Charter of the Association of Southeast Asian Nations, Singapore, November 20, 2007.

Charter of the Cooperation Council for the Arab States of the Gulf, Abu Dhabi, United Arab Emirates, May 25, 1981. As of December 2, 2009:
http://www.gccsg.org/eng/index.php?action=Sec-Show&ID=1

Charter of the Organisation of the Islamic Conference, Dakar, Senegal, March 14, 2008.

Conference on Security and Co-operation in Europe: Final Act, Helsinki, Finland, August 1, 1975.

Convention of the Organisation of the Islamic Conference on Combating International Terrorism, 1999. As of November 15, 2009:
http://www.oicun.org/print.php?item_id=&sec_id=38

Declaration of Principles on Interim Self-Government Arrangements, Washington, D.C., September 13, 1993.

India-Russia Joint Declaration, Moscow, Russia, December 7, 2009.

Interim Agreement Between the United States of America and the Union of Soviet Socialist Republics on Certain Measures with Respect to the Limitation of Strategic Offensive Arms, Moscow, Soviet Union, May 26, 1972.

Istanbul Cooperation Initiative, North Atlantic Treaty Organization, June 28, 2004. As of December 2, 2009:
http://www.nato.int/docu/comm/2004/06-istanbul/docu-cooperation.htm

Joint Declaration of the Paris Summit for the Mediterranean, Paris, France, July 2008.

Memorandum of Understanding Between the Government of the United States of America and the Government of the Union of Soviet Socialist Republics Regarding the Establishment of a Standing Consultative Commission, Geneva, Switzerland, December 21, 1972.

Modified Brussels Treaty, Paris, France, October 23, 1954.

Mutual Defence Assistance Agreement, Pakistan and the United States of America, Karachi, Pakistan, May 19, 1954.

North Atlantic Treaty, Washington, D.C., April 4, 1949.

Partnership for Peace: Framework Document, issued by the heads of state and government participating in the meeting of the North Atlantic Council, January 10–11, 1994. As of December 2, 2009:
http://www.nato.int/cps/en/SID-A5EE13AE-1CB89703/natolive/official_texts_24469.htm?selectedLocale=en

Petersberg Declaration, Western European Union Council of Ministers, Bonn, Germany, June 19, 1992.

Resolution on Islamic Office for the Boycott of Israel, 36th Session of the Council of Foreign Ministers, Damascus, Syria, May 23–25, 2009.

Riga Summit Declaration, North Atlantic Treaty Organization, November 29, 2006.

Second Protocol Amending the Treaty of Amity and Cooperation in Southeast Asia, Manila, Philippines, July 25, 1998.

Summit Declaration on Afghanistan, North Atlantic Treaty Organization, April 4, 2009.

Treaty of Amity and Cooperation in Southeast Asia, Indonesia, February 24, 1976. As of November 15, 2009:
http://www.aseansec.org/1217.htm

Treaty of Economic, Social, and Cultural Collaboration and Collective Self-Defense, Brussels, Belgium, March 17, 1948.

Treaty of Friendship, Cooperation and Mutual Assistance Between the People's Republic of Albania, the People's Republic of Bulgaria, the Hungarian People's Republic, the German Democratic Republic, the Polish People's Republic, the Rumanian People's Republic, the Union of Soviet Socialist Republics and the Czechoslovak Republic, May 14, 1955.

Treaty of Peace Between the State of Israel and the Hashemite Kingdom of Jordan, October 26, 1994.

United Nations Security Council, Resolution 619, S/RES/619(1988), August 9, 1988.

———, Resolution 686, S/RES/0686(1991), March 2, 1991.

———, Resolution 687, S/RES/0687(1991), April 3, 1991.

———, Resolution 688, S/RES/0688(1991), April 5, 1991.

———, Resolution 689, S/RES/0689(1991), April 9, 1991.

———, Resolution 692, S/RES/0692(1991), May 20, 1991.

———, Resolution 700, S/RES/0700(1991), June 17, 1991.

———, Resolution 705, S/RES/0705(1991), August 15, 1991.

———, Resolution 706, S/RES/0706(1991), August 15, 1991.

———, Resolution 707, S/RES/0707(1991), August 15, 1991.

———, Resolution 712, S/RES/0712(1991), September 19, 1991.

———, Resolution 715, S/RES/0715(1991), October 11, 1991.

———, Resolution 984, S/RES/0984(1995), April 11, 1995.

———, Resolution 986, S/RES/0986(1995), April 14, 1995.